SpringerBriefs in Environmental Science

SpringerBriefs in Environmental Science present concise summaries of cutting-edge research and practical applications across a wide spectrum of environmental fields, with fast turnaround time to publication. Featuring compact volumes of 50 to 125 pages, the series covers a range of content from professional to academic. Monographs of new material are considered for the SpringerBriefs in Environmental Science series.

Typical topics might include: a timely report of state-of-the-art analytical techniques, a bridge between new research results, as published in journal articles and a contextual literature review, a snapshot of a hot or emerging topic, an in-depth case study or technical example, a presentation of core concepts that students must understand in order to make independent contributions, best practices or protocols to be followed, a series of short case studies/debates highlighting a specific angle.

SpringerBriefs in Environmental Science allow authors to present their ideas and readers to absorb them with minimal time investment. Both solicited and unsolicited manuscripts are considered for publication.

Mandy A. van den Ende • Arjan Wardekker
Dries L. T. Hegger • Heleen L. P. Mees
Joost M. Vervoort

Towards a Climate-Resilient Future Together

Tools for Engaging Citizens for a Better Future

Springer

Mandy A. van den Ende
Copernicus Institute of Sustainable Development
Utrecht University
Utrecht, The Netherlands

Arjan Wardekker
Centre for the Study of the Sciences and the Humanities
University of Bergen
Bergen, Norway

Dries L. T. Hegger
Copernicus Institute of Sustainable Development
Utrecht University
Utrecht, The Netherlands

Heleen L. P. Mees
Copernicus Institute of Sustainable Development
Utrecht University
Utrecht, The Netherlands

Joost M. Vervoort
Copernicus Institute of Sustainable Development
Utrecht University
Utrecht, The Netherlands

Utrecht University (Water, Climate & Future Deltas hub)

ISSN 2191-5547 ISSN 2191-5555 (electronic)
SpringerBriefs in Environmental Science
ISBN 978-3-031-07681-7 ISBN 978-3-031-07682-4 (eBook)
https://doi.org/10.1007/978-3-031-07682-4

This Springer imprint is published by the registered company Springer Nature Switzerland AG
The registered company address is: Gewerbestrasse 11, 6330 Cham, Switzerland

Preface

1997: The most recent edition of the 'Elfstedentocht'[1]
2020: One of the longest heatwaves since 1901
2020: Long periods of droughts
2021: Extreme river flooding in the province of Limburg; the second most expensive natural disaster of 2021 globally
2022: Four storms in less than 1 month
Throughout the years: pluvial floods

These are only a few examples of climate change impacts experienced in a relatively small country: the Netherlands. Imagine this list if it would include *all* climate change impacts *globally*. We would probably be able to fill an entire library.

The increased intensity and frequency of climate change impacts make that, more than ever in modern history, the future well-being and even survival of human communities all around the world depend on their ability to become climate proof or *resilient*. This demands joint climate adaptation efforts by governments, companies and citizens. In previous research, we, the authors, found that the latter group is often being overlooked in climate adaptation endeavours. A missed opportunity, given that citizens (1) can improve the quality of new adaptation policies and plans with their holistic, locally grounded perspectives; (2) are needed to support the implementation of new adaptation policies and plans; and (3) can implement adaptation measures themselves. Besides, citizens are the ones living in these communities, so should they just not be given the chance to design, plan and build their own future living environment? Many local governments do underline the importance of involving citizens in making a place climate resilient, yet organizing this in a meaningful way remains challenging. Challenges can partly be attributed to the focus on 'the future', which is rather abstract to people, citizens in particular. As the distinguished Dutch professor of urban futures Maarten Hajer would put it: there is a need to better mobilize the power of imagination. We believe that 'foresight' or 'futures' processes have a valuable contribution to make here. While much has been written

[1] A Dutch tradition: a 200 kilometres ice skating tour on natural ice.

on the technical aspects of scenario methods, there is little practical guidance on what might make it engaging to citizens. Rather than recruiting citizens into what feels like a technical process, an actual collaboration is required.

This toolbox entitled *Towards a Climate-Resilient Future Together* collects and showcases some of the lessons learned from several international research programmes on citizen engagement in the form of practical foresight methods and step-by-step advice on how to apply them. These programmes include CoCliServ (Co-development of Place-Based Climate Services for Action; funded by EU JPI Climate/ERA4CS), CCAFS (Climate Change, Agriculture and Food Security; funded by CGIAR global research partnership), and Utrecht University's Water, Climate and Future Deltas hub. The latter funded the development of this toolbox. By offering a complete set of practical foresight methods, tools and tips, the toolbox can be seen as a welcome companion for policymakers, consultants or researchers that find themselves in the process of involving citizens in thinking about and developing a climate-resilient future, *together!*

Utrecht, The Netherlands — Mandy van den Ende
March 2022

Acknowledgements

This toolbox has been developed as part of the international research project CoCliServ (funded by ERA4CS, JPI Climate). Our special thanks go to Utrecht University's Water, Climate and Future Deltas hub, co-funder of this toolbox. We would also like to thank the Municipality of Dordrecht, Studio Lakmoes, CAS Climate Adaptation Services, KNMI Netherlands Meteorological Institute, Vogelnest neighbourhood centre and residents of the Vogelbuurt neighbourhood for collaborating in the workshop that illustrates the methods presented in this toolbox. Finally, Margot Stoete is thanked for her help with some of the visuals in this book.

Utrecht, The Netherlands — Mandy van den Ende

Bergen, Norway — Arjan Wardekker

Utrecht, The Netherlands — Heleen Mees

Utrecht, The Netherlands — Dries Hegger

Utrecht, The Netherlands — Joost Vervoort
March 2022

Contents

1 Introduction to the Participatory Foresight Toolbox 1
1.1 Why This Toolbox? 4
1.2 Who Can Use This Toolbox? 4
References 5

2 Preparing Participatory Foresight Methods 7
2.1 Why Citizen Participation? 8
2.1.1 Participation Goal 1: Policy Development 9
2.1.2 Participation Goal 2: Community Building 13
2.1.3 Participation Goal 3: Knowledge and Capacity Building . . . 14
2.2 Who to Involve? 14
2.2.1 Stakeholders and Participants 15
2.2.2 How to Motivate People to Participate? 16
2.3 Which Foresight Methods to Use? 19
2.3.1 Exploratory Scenarios 21
2.3.2 Visions of a Desired Future 22
2.3.3 Pathways to the Desired Future 24
2.3.4 Combinations of Methods in the Same Activity 28
Appendix A 30
Additional Information About Tools for Preparing
Participatory Foresight Activities 30
References 33

3 Performing Participatory Foresight Methods 37
3.1 Exploratory Scenarios 38
3.1.1 How to Use Exploratory Scenarios
in a Participatory Activity? 38
3.2 Visioning 39
3.2.1 How to Use Visioning in a Participatory Activity? 39
3.2.2 Output Examples 41

3.3 Backcasting 43
3.3.1 How to Use Backcasting in a Participatory Activity? 43
3.3.2 Output Examples 46
3.4 Incremental Backcasting 47
3.4.1 How to Use Incremental Backcasting in a Participatory Activity? 47
3.4.2 Output Examples 47
3.5 Seeds-Based Pathways 48
3.5.1 How to Use Seeds-Based Pathways in a Participatory Activity? 48
3.5.2 Output Examples 48
Appendix A 51
Additional Information about Tools for Performing Participatory Foresight Activities 51
Appendix B: Example List of Climate Adaptation Measures in Urban Areas 57
References 58

4 Reflection: Applying Participatory Foresight Methods in Practice 61
4.1 A Participatory Foresight Workshop in Dordrecht, the Netherlands (Urban Context) 61
4.1.1 Why Citizen Participation? 62
4.1.2 Who to Involve? 62
4.1.3 Which Foresight Methods to Use? And How to Use Foresight Methods? 63
4.2 A Participatory Foresight Workshop in Honduras (Rural Context) 66
4.2.1 Why Citizen Participation? 67
4.2.2 Who to Involve? 67
4.2.3 Which Foresight Methods to Use? And How to Use Foresight Methods? 68
References 70

Chapter 1
Introduction to the Participatory Foresight Toolbox

Abstract We start this first chapter of the participatory foresight toolbox with an introduction to 'climate resilience'. We, next, argue for the importance of a climate-resilient living environment given frequent and severe climate change impacts. We then explain why citizens play a crucial role in contributing to such a climate-resilient future and we propose 'foresight' as a suitable methodological approach for engaging them herein. In the conclusion of this chapter, we sum up why this toolbox is a welcome companion for organizers of participatory foresight activities.

Keywords Toolbox · Citizen participation · Futures · Foresight methods · Scenarios · Visioning · Pathways · Resilience · Climate adaptation · Urban · Rural · Sustainability

Urban and rural communities worldwide face a wide array of complex challenges at once when it comes to spatial development. On the one hand, many of them experience severe climate change impacts such as floods, heat stress, and droughts, which are projected to happen more frequently in the future (IPCC 2018). In light of future climate change, it is crucial for communities to not only mitigate but also adapt to these climate change impacts, hence increasing their *resilience* against shocks and disturbances. Climate resilience can be defined as "the ability [of a city or urban system] to absorb disturbance while retaining identity, structure and key processes" (Leichenko 2011, p. 164). In practice, however, people have a wide variety of perspectives on what climate resilience means, and how a 'climate-resilient future' should be designed, prioritised and built (Wardekker 2021). Besides climate change, many communities also face various socioeconomic issues concerning spatial development, such as housing demand and degrading neighbourhoods (Wardekker et al. 2020).

A way to comprehensively deal with societal challenges, among which climate adaptation, is to use a sustainable development approach (UN Environment 2019). Following this approach, spatial development policy and plans ideally benefit the biosphere, society and economy simultaneously (Fig. 1.1).

M. A. van den Ende et al., *Towards a Climate-Resilient Future Together*, SpringerBriefs in Environmental Science,
https://doi.org/10.1007/978-3-031-07682-4_1

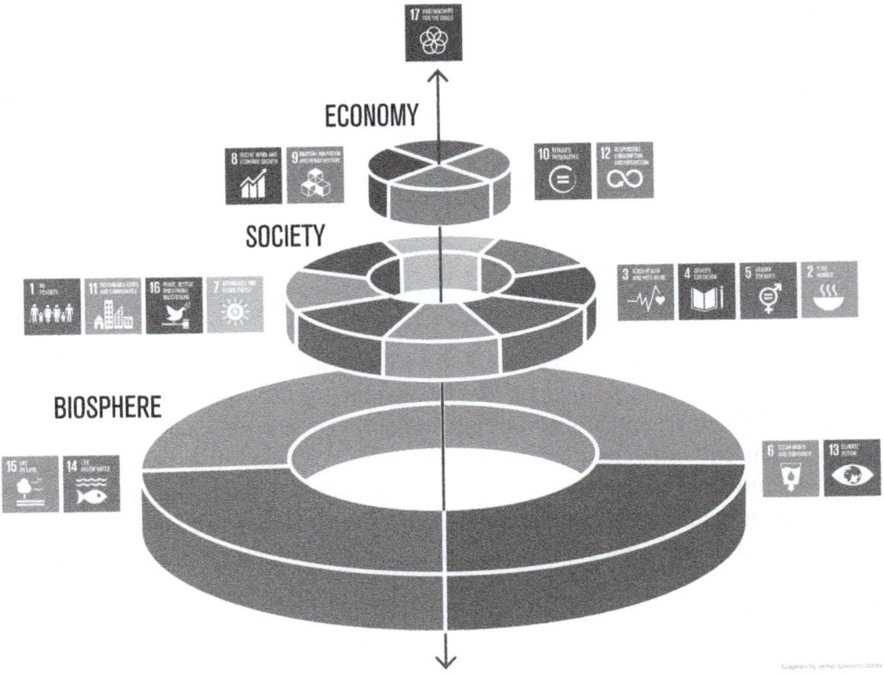

Fig. 1.1 The embeddedness of economies, societies and the biosphere. (From Azote for Stockholm Resilience Centre, Stockholm University)

However, in practice, climate adaptation in the context of spatial development is usually not very focused on 'social' aspects (Lorenz et al. 2017; Wardekker et al. 2020); it often concerns technological measures, such as strengthening dikes and dams or renewing sewages. Citizens are typically excluded from these policymaking processes, leading to less engagement with the issue of climate adaptation in general. Various studies, however, emphasize their important role in climate adaptation (Mees et al. 2019; Uittenbroek et al. 2019b). For instance, a recent study found that citizens perceive climate adaptation more holistically; they typically touch upon various dimensions of sustainable development as compared to representatives of institutional actors (Marschütz et al. 2020). These multi-issue perspectives can form the basis of more innovative climate risk management approaches that simultaneously contribute to other priorities, such as improving the quality of the neighbourhood (Wardekker et al. 2020). This way, citizens can improve the quality of knowledge used in climate adaptation (Bremer et al. 2021). Similar studies show that citizens feel their wishes, desires and fears are neglected when climate adaptation plans are imposed while they have to live with these plans, as well as bear the implementation costs. Hence, citizen engagement could legitimize climate adaptation action (Marschütz et al. 2020; Uittenbroek et al. 2019a). Another reason for more citizen engagement in policy processes is to empower them as well as to encourage climate adaptation actions and initiatives, such as installing green roofs, placing rain barrels or removing pavement from gardens (Brink and Wamsler 2018; Hegger et al. 2017; Mees et al. 2013; Uittenbroek et al. 2019a). In sum, citizen involvement is essential for a climate-resilient future, because:

1. Citizens can enrich climate adaptation plans with their holistic priorities;
2. Citizens are needed to support the implementation of climate adaptation plans; and
3. Citizens can implement climate adaptation measures themselves

Besides citizen involvement, sustainable development also requires developing actions in the short-term in such a way that they contribute to long-term sustainability. This is where the role of foresight comes in, simply defined as "the act of thinking about the future to guide decisions today" (Wiebe et al. 2018, p. 546). If there was just one path to a sustainable future, developing sustainable measures would be a relatively simple task. However, we live in a complex, dynamic world with climate disturbances occurring at unexpected moments in time, with varying levels of intensity and at different geographical scales. Due to these uncertainties, there are multiple 'futures' possible (Haasnoot et al. 2013). Predicting one most likely future (Fig. 1.2, left) as a basis for policymaking on sustainability problems such as climate adaptation is, therefore, not very useful. Instead, to anticipate uncertainty it is important to take a broader perspective and consider multiple different possible futures, including more desired futures (Fig. 1.2, right). This way, new insights about opportunities and challenges that different futures may bring can be considered and acted upon in the present.

Dealing with uncertain futures requires actors to imagine situations that are completely different from the present. Methodologically speaking, it helps to use creative foresight methods and tools. Furthermore, the inclusion of different people such as policymakers and citizens can increase the diversity of perspectives.

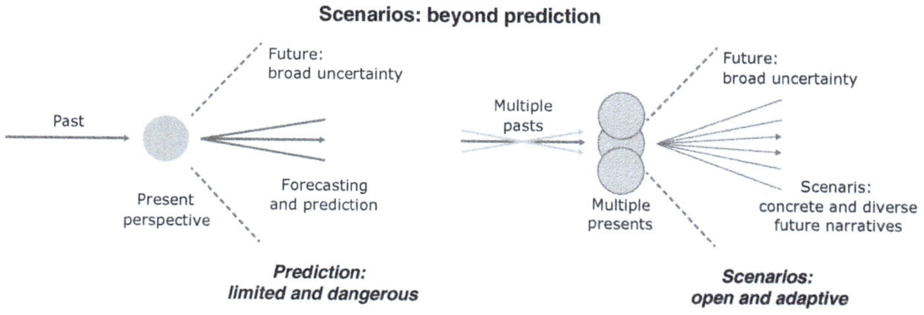

Fig. 1.2 Predicting one future (left) versus anticipating multiple futures (right). (Adapted from Noortmann et al., 2019)

1.1 Why This Toolbox?

In the previous paragraphs, we have discussed that there is an apparent need for (1) more inclusive, democratic approaches to sustainable development. We need to move beyond top-down approaches to climate action and engage citizens more actively in decisions that concern their own living environment. Although many local governments may have formulated ambitions to involve the wider public in policy processes, they often struggle with organizing meaningful participatory activities (Uittenbroek et al. 2019b). The result – distrust and dissatisfaction among citizens – is something we want to avoid with this toolbox. We also identified the need for (2) guidance on how to deal with complex local challenges in light of an uncertain future, such as making a living environment climate-resilient. What is missing is a step-by-step guide for using the various methods collectively described as 'foresight'. Foresight methods include, among other methods, *visioning* and *backcasting*, which lend themselves to constructing ideal climate-resilient futures and formulating ways to achieve them (Mangnus et al. 2019). Another foresight method, *exploratory scenarios*, can be used to 'crash test' these desired visions of the future (Shaw et al. 2009; Sheppard et al. 2011; Vervoort et al. 2014). This toolbox gives hands-on tips and tricks on how to organize participatory foresight activities with citizens.

1.2 Who Can Use This Toolbox?

The toolbox is developed for practitioners: local governments (e.g. policymakers), NGOs and community leaders/organizers, and third parties (e.g. consultants), who want to engage citizens in thinking about and contributing to a climate-resilient future. The step by step guidance makes it suitable for readers without any experience in citizen participation and/or foresight whatsoever. At the same time, more experienced readers will find innovative combinations of methods and tools that are unique within the field of citizen participation and foresight. Interested readers will be given more detailed background information as well as be updated with the latest case study examples.

The toolbox is structured in such a way that it guides the reader through the process of organizing participatory foresight activities. In the 'preparation phase' (Chap. 2), we discuss important matters to consider when organizing activities that involve citizens. This starts by inviting the organizer to critically think about the question of *why citizen participation* (Sect. 2.1). This goal is important as it closely relates to the questions of *who to involve* (Sect. 2.2) and *which foresight methods to use* (Sect. 2.3). Having prepared the participatory activity, it is time for the 'action phase' (Chap. 3), in which we explain *how to use foresight methods* in practice. In Chap. 4, we reflect on the toolbox by applying Chaps. 2 and 3 to real-time examples of participatory foresight activities: one workshop organized in an urban delta in The Netherlands and one workshop in a rural area in Africa.

©Studio Lakmoes

References

Bremer S, Wardekker A, Jensen ES, van der Sluijs JP (2021) Quality assessment in co-developing climate services in Norway and The Netherlands. Front Clim 3:627665. https://doi.org/10.3389/fclim.2021.627665

Brink E, Wamsler C (2018) Collaborative governance for climate change adaptation: mapping citizen-municipality interactions. Environ Policy Gov 28:82–97. https://doi.org/10.1002/eet.1795

Haasnoot M, Kwakkel JH, Walker WE, Ter Maat J (2013) Dynamic adaptive policy pathways: a method for crafting robust decisions for a deeply uncertain world. Glob Environ Chang 23:485–498. https://doi.org/10.1016/j.gloenvcha.2012.12.006

Hegger DLT, Mees HLP, Driessen PPJ, Runhaar HAC (2017) The roles of residents in climate adaptation: a systematic review in the case of the Netherlands. Environ Policy Gov 27:336–350. https://doi.org/10.1002/eet.1766

IPCC (2018) Summary for policymakers. In: Global warming of 1.5°C. an IPCC special report on the impacts of global warming of 1.5°C above pre-industrial levels and related global greenhouse gas emission pathways, in the context of strengthening the global response to. https://www.ipcc.ch/

Leichenko R (2011) Climate change and urban resilience. Environ Sustain 3(3):164–168. https://doi.org/10.1016/j.cosust.2010.12.014

Lorenz S, Dessai S, Forster PM, Paavola J (2017) Adaptation planning and the use of climate change projections in local government in England and Germany. Reg Environ Chang 17:425–435. https://doi.org/10.1007/s10113-016-1030-3

Mangnus AC, Vervoort JM, Mcgreevy SR, Ota K, Rupprecht CDD, Oga M, Kobayashi M (2019) New pathways for governing food system transformations: a pluralistic practice-based futures approach using visioning, back-casting, and serious gaming. Ecol Soc 24(4):2. https://doi.org/10.5751/ES-11014-240402

Marschütz B, Bremer S, Runhaar H, Hegger D, Mees H, Vervoort J, Wardekker A (2020) Local narratives of change as an entry point for building urban climate resilience. Clim Risk Manag 28:100223. https://doi.org/10.1016/j.crm.2020.100223

Mees HLP, Driessen PPJ, Runhaar HAC, Stamatelos J (2013) Who governs climate adaptation? Getting green roofs for stormwater retention off the ground. J Environ Plan Manag 56(6):802–825. https://doi.org/10.1080/09640568.2012.706600

Mees HLP, Uittenbroek CJ, Hegger DLT, Driessen PPJ (2019) From citizen participation to government participation: an exploration of the roles of local governments in community initiatives for climate change adaptation in the Netherlands. Environ Policy Gov 29:198–208. https://doi.org/10.1002/eet.1847

Noortmann M, Koning J, Vervoort J, Hoofd I (2019) Imaginative scenario planning for security and law enforcement organisations. A Report on practicing with uncertain security futures

Shaw A, Sheppard S, Burch S, Flanders D, Wiek A, Carmichael J, Robinson J, Cohen S (2009) Making local futures tangible—synthesizing, downscaling, and visualizing climate change scenarios for participatory capacity building. Glob Environ Chang 19:447–463. https://doi.org/10.1016/j.gloenvcha.2009.04.002

Sheppard SRJ, Shaw A, Flanders D, Burch S, Wiek A, Carmichael J, Robinson J, Cohen S (2011) Future visioning of local climate change: a framework for community engagement and planning with scenarios and visualisation. Futures 43:400–412. https://doi.org/10.1016/j.futures.2011.01.009

Uittenbroek CJ, Mees HLP, Hegger DLT, Driessen PPJ (2019a) From public to citizen responsibilities in urban climate adaptation: a thick analysis. In: Urban climate politics: agency and empowerment, pp 171–189. https://doi.org/10.1017/9781108632157.010

Uittenbroek CJ, Mees HLP, Hegger DLT, Driessen PPJ (2019b) The design of public participation: who participates, when and how? Insights in climate adaptation planning from the Netherlands. J Environ Plan Manag 62(14):2529–2547. https://doi.org/10.1080/09640568.2019.1569503

UN Environment (2019) Global environment outlook – GEO-6: summary for policymakers. https://doi.org/10.1017/9781108639217

Vervoort JM, Thornton PK, Kristjanson P, Förch W, Ericksen PJ, Kok K, Ingram JSI, Herrero M, Palazzo A, Helfgott AES, Wilkinson A, Havlík P, Mason-D'Croz D, Jost C (2014) Challenges to scenario-guided adaptive action on food security under climate change. Glob Environ Chang 28:383–394. https://doi.org/10.1016/j.gloenvcha.2014.03.001

Wardekker A (2021) Contrasting the framing of urban climate resilience. Sustain Cities Soc 75:103258. https://doi.org/10.1016/j.scs.2021.103258

Wardekker A, Van den Ende M, Marschutz B, Pijnappels M, Hofland S, Bremer S, Blanchard A, Iversen L, Van der Sluis J, Van Krauß W, Rocha A, Da Cunha C, Baztan J, Jaffrès L (2020) Incremental scenario case studies. CoCliServ, Guyancourt, France.

Wiebe K, Zurek M, Lord S, Brzezina N, Gabrielyan G, Libertini J, Loch A, Thapa-Parajuli R, Vervoort J, Westhoek H (2018) Scenario development and foresight analysis: exploring options to inform choices. Annu Rev Environ Resour 43:545–570. https://doi.org/10.1146/annurev-environ-102017-030109

Chapter 2
Preparing Participatory Foresight Methods

Abstract This chapter is about the process of preparing participatory foresight methods (see Fig. 2.1). In Sect. 2.1 (*Why citizen participation?*), we discuss different goals of citizen participation in foresight activities: (I) policy development; (II) community building; and (III) knowledge and capacity building. In Sect. 2.2 (*Who to involve?*), we focus on participants and provide practical tools to encourage their involvement in the foresight activity. Finally, in Sect. 2.3 (*Which foresight methods to use?*), we introduce different foresight methods and discuss their relevance for the three goals of citizen participation. We distinguish between three types of foresight methods: (1) exploratory scenarios, to explore a range of plausible futures; (2) visioning, to imagine a desired climate-resilient future; and (3) pathways, to explore routes to that desired climate-resilient future.

Keywords Toolbox · Citizen participation · Futures · Foresight methods · Scenarios · Visioning · Pathways · Resilience · Climate adaptation · Urban · Rural · Sustainability

M. A. van den Ende et al., *Towards a Climate-Resilient Future Together*, SpringerBriefs in Environmental Science,
https://doi.org/10.1007/978-3-031-07682-4_2

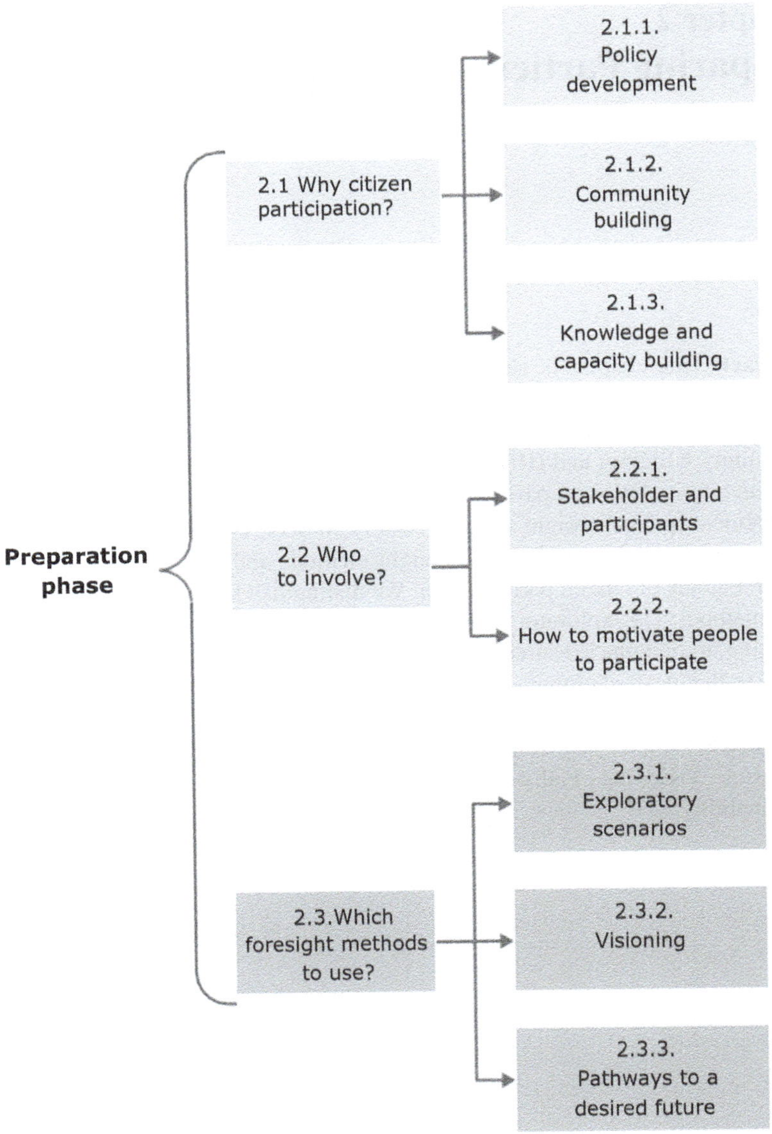

Fig. 2.1 Overview of this chapter

2.1 Why Citizen Participation?

Participatory methods require investments (e.g. time, money, efforts) from both participants and organizers. The first important step in the preparation phase is, therefore, to critically think about the goal of citizen participation (Uittenbroek et al. 2019). Based on the literature, we assume organizers to pursue one or more of the following goals of citizen participation in foresight activities (see Fig. 2.2):

1. **Policy development:** to inform local climate policy and plans with the input of citizens (see Sect. 2.1.1);
2. **Community building:** to stimulate learning about climate change, raise awareness, and encourage citizen action (see Sect. 2.1.2);
3. **Knowledge and capacity building:** to identify the type of information or support citizens need to adapt to climate change (see Sect. 2.1.3).

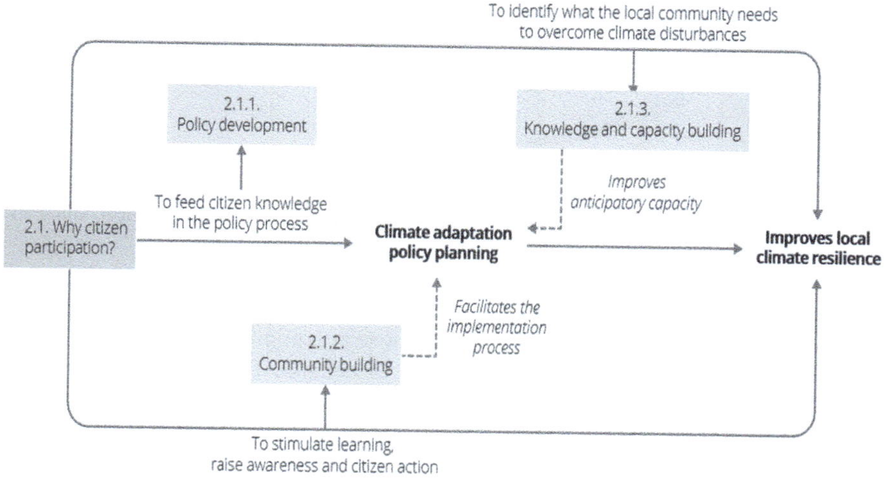

Fig. 2.2 Overview of Sect. 2.1

Note 1: *The chosen goal of citizen participation influences, in turn, other choices such as who to involve (Sect. 2.2) and which foresight methods to use (Sect. 2.3)* (Uittenbroek et al. 2019).

Note 2: *It can be disappointing for citizens if they expect their input to influence the policy process, while project organizers only aimed at raising awareness about climate change or encouraging citizen action. Especially in a time where people's trust in public institutions has become an increasing issue of concern, it is essential to be transparent about the intended goal of participation beforehand* (Ramos et al. 2019).

2.1.1 Participation Goal 1: Policy Development

Da Costa et al. (2008, p. 369) define the goal of using citizen input to 'inform policy' as: "generating insights regarding the dynamic of change, future challenges and options, along with new ideas, and transmitting them to policymakers as an input to policy conceptualisation and design." From the perspective of policymakers, citizen input, which is typically more socially relevant, experimental, and robust, can be

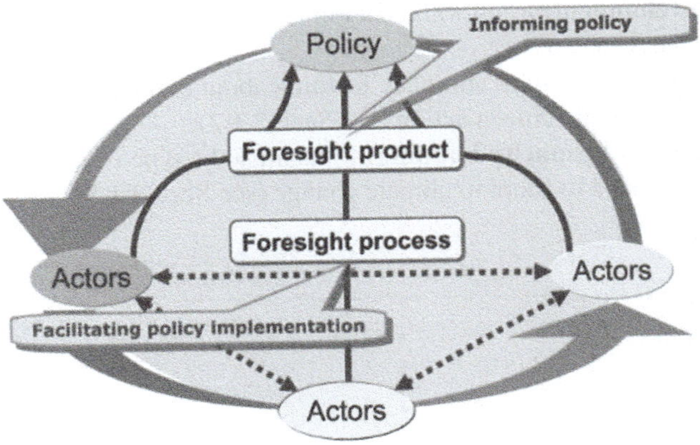

Fig. 2.3 Citizen participation in foresight activities for policy development: to inform policy and to facilitate policy implementation. (From Da Costa et al. 2008)

seen as the *product* of participatory foresight activities (see Fig. 2.3). At the same time, by taking the desires and concerns of those potentially affected into account, the *process* of citizen participation is likely to enhance the legitimacy of policy plans, hence facilitating policy implementation (Da Costa et al. 2008; Marschütz et al. 2020).

Note 3: *If policy development is the intended goal of citizen participation, it requires consideration to what extent participants can indeed influence decisions* (Uittenbroek et al. 2019). *See Sect. 2.1.1.1 for more information on the degree of citizen participation.*

2.1.1.1 The Degree of Citizen Participation

Participation processes can thus improve the effectiveness and legitimacy of climate adaptation plans (Da Costa et al. 2008; Marschütz et al. 2020; Uittenbroek et al. 2019). However, the degree of influence (also referred to as the 'degree of participation') on local adaptation plans can vary, often depending on *when* citizens are involved in the policy process (Uittenbroek et al. 2019). For instance, they can provide input in an early phase of agenda-setting, but they can also be consulted to respond to already developed plans. Therefore, to think about the degree of participation, we use the participation ladder of Arnstein (1969) (see Fig. 2.4) in combination with the policy cycle (see Fig. 2.5) (Vaz and Prendeville 2019).

The top of the participation ladder shows the highest degrees of participation, which are seen as forms of citizen empowerment (Arnstein 1969). To empower citizens in policy making, it is important to involve them in early policy phases (e.g. the problem identification, agenda-setting, or policy formulation phase) to avoid that "public opinion usually enters into the planning process at such a late stage as to have minor influence on actual outcomes" (Carlsson-Kanyama et al. 2008, p. 44).

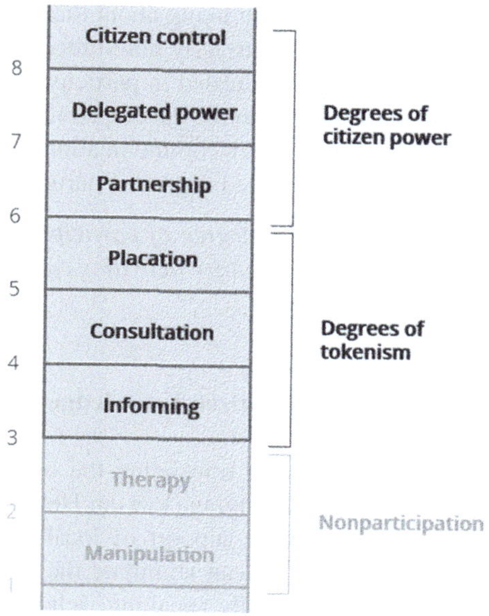

Fig. 2.4 Participation ladder [adjusted]. (Adapted from Arnstein 1969)

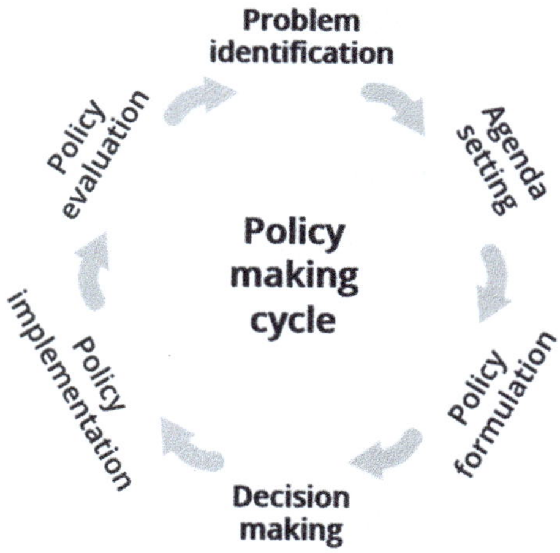

Fig. 2.5 The policy cycle. (Adapted from Vaz and Prendeville 2019)

Also note that high degrees of participation require policymakers to be flexible and open to different visions that may steer their initial ideas in alternative directions.

Citizens can also inform, consult or advice local decisions about climate adaptation. These forms of participation are lower on the participation ladder; citizens can at most 'respond' to plans. Organizers may choose a low degree of participation if there is limited room to manoeuvre, for instance, because they have to deliver certain outcomes. Although Arnstein (1969) calls low degrees of participation 'forms of tokenism', in some cases less deliberation of citizen knowledge is not necessarily problematic (Uittenbroek et al. 2019). It is more a question of *who* to involve *when*. For instance, in the agenda-setting phase, it may be decided to involve a diverse group of citizens in thinking about their future living environment, while in a later design phase, only strongly committed citizens or those with legal and technical expertise may be interested in participation. A broader group of citizens can again be asked to respond to the developed (draft) plans (Wardekker et al. 2020b). In any case, it requires good communication, openness and transparency towards citizens about the intended degree of participation.

Note 4: *For a high degree of participation, it requires good thinking on how the output of participatory activities can be translated into actual policy outcomes (see Sect. 2.1.1.2).*

2.1.1.2 Feeding Citizen Knowledge in Policy

Research shows that translating the output of participatory activities into policy action is often challenging (Van der Hel and Biermann 2017; Vervoort and Mangnus 2018). Although the support of local policymakers cannot be enforced, several enabling conditions can stimulate the uptake of proposed visions and actions in policymaking. Broadly speaking, scholars emphasize the importance of political authority – that is, when local policymakers perceive the participatory activity as *salient*, *credible* and *legitimate*. Communication, translation and mediation are hereby crucial (Cash et al. 2003; Van der Hel and Biermann 2017).

Communication

One potential obstacle with regards to producing salient output is that citizens and policymakers often communicate from highly different views. For example, citizens may envision a radically different future neighbourhood full of innovative climate adaptation measures, whereas policymakers may prefer to work with what they know or are bound by municipal rules and policy (Bahadur and Tanner 2014; Van den Ende et al. 2022). The use of multiple foresight methods can help to make the output 'true to life'. For instance, one can expand on general visions of a climate-resilient future by formulating feasible short and medium-term actions to get there.

Another challenge is that the output of participatory activities is easily ignored if not in line with the political agenda or interests of policymakers. One way to better communicate the value of the output is to invite policymakers to attend the participatory activity themselves (Bremer et al. 2021; Cash et al. 2003; Dinesh et al. 2021), or by linking the participatory activity to a specific event in time, i.e. a window of opportunity (Wardekker et al. 2020b). In the context of this toolbox, planned urban renewal projects such as the reconstruction of a sewage system, new houses that will be built or the renovation of public space are good moments to integrate the input of citizens in the policy process from the start.

Finally, the basis of good communication is mutual trust between citizens and policymakers, which often is lacking (Crawford 2019; Wardekker et al. 2020b). In the first place, it is essential that policymakers are willing to extend the notion of 'expert' and consider citizen knowledge legitimate input for policymaking (Lorenz et al. 2017; Uittenbroek et al. 2019).

Translation

Policymakers and citizens often speak in different 'languages'. Mutual understanding of events or phenomena can be facilitated by translating information in a way that both citizens and policymakers perceive as relevant, credible and legitimate (Cash et al. 2003; Da Costa et al. 2008). Among others, visualization can help understand complex climate information and stimulate learning among participants. For instance, visualizations of the local landscape link to people's attachment to place and community identity and thereby bridge the gap between formal models and local realities (Sheppard et al. 2011; Wardekker et al. 2020b).

Mediation

Mediation by an external party can also contribute to a better uptake of citizen knowledge in policy. External facilitators are usually in a better position to find overlap in perspectives and bridge interests (Cash et al. 2003). They can also plan follow-up activities to ensure the participatory activities find some institutionalization in the policy process. The preferred role of facilitators depends on the goal of participation. For instance, facilitators may encourage out-of-the-box thinking if the goal is to enhance awareness about possible futures (Wardekker et al. 2020b). However, more specific output may be preferred when the goal is to inform policy. If facilitators are completely non-interventionistic here, allowing groups to discuss freely rather than to come to the point, the output will likely be less concrete, hence less relevant for policymakers.

2.1.2 Participation Goal 2: Community Building

Another possible goal of organizing participatory foresight activities is 'community building', which refers to generating societal impact (Da Costa et al. 2008; Crawford 2019). In the context of this toolbox, societal impact would be preparing citizens for climate change impacts by raising awareness and stimulating their engagement in the issue of climate adaptation. The participatory activity, in this regard, can be seen as a:

> 'learning and experience space,' in which knowledge of the social-ecological crisis is shared, a space that is not limited to biophysical knowledge, but also includes 'underlying knowledge of the basic causes (where our system fails), of possible strategies for change, and of possible alternatives (able/daring to think of a different world)' (Peeters 2014, p. 99)

The idea is that people first need to realize that climate change is also *their* problem to be able to reflect on their own behaviour and ability to live with climate change impacts. For example, participatory foresight activities can make citizens aware that the impact of a cloudburst can be reduced if many of them would regreen their gardens or install green roofs for rainwater infiltration and storage capacity (Hegger et al. 2017; Mees et al. 2013). Also social cohesion can contribute to climate resilience, for instance, if citizens would give each other a helping hand in times of heat stress or floods (Hegger et al. 2017). Participatory foresight activities can also encourage collective action, for instance, citizens initiating community gardens. Organized groups of citizens also have more power to influence policy-making in line with community priorities (Tosun and Schoenefeld 2017). Hence, participatory foresight activities aiming at community building can enhance local climate resilience.

2.1.3 *Participation Goal 3: Knowledge and Capacity Building*

The third goal of citizen participation in foresight activities, 'knowledge and capacity building', refers to identifying what the local community needs to adapt to climate change. These 'climate services' can range from information on what their living environment could look like under different scenarios, to cost-benefit calculations of concrete climate action plans, or better communication at times when it is especially needed. While knowledge and awareness is not the only issue driving local interest in adaptation – citizens may simply have other pressing concerns that take priority – it is part of the equation (Wardekker et al. 2020b). By tailoring climate services to the local situation, citizens and other local actors are better able to weigh the local relevance of climate change impacts, policy options, their own roles in adaptation, and how it all might tie in with other concerns (Wardekker et al. 2020a, b). Thereby, participatory foresight activities aiming at knowledge and capacity building could contribute to the overall goal of building local climate resilience.

2.2 Who to Involve?

After having decided on the goal of citizen participation, the next step is to think about who to involve in the participatory foresight activity. The current section helps identify relevant stakeholders and discusses the questions of which participants to invite when (Sect. 2.2.1) and how to encourage citizens to participate in foresight activities (Sect. 2.2.2) (see Fig. 2.6).

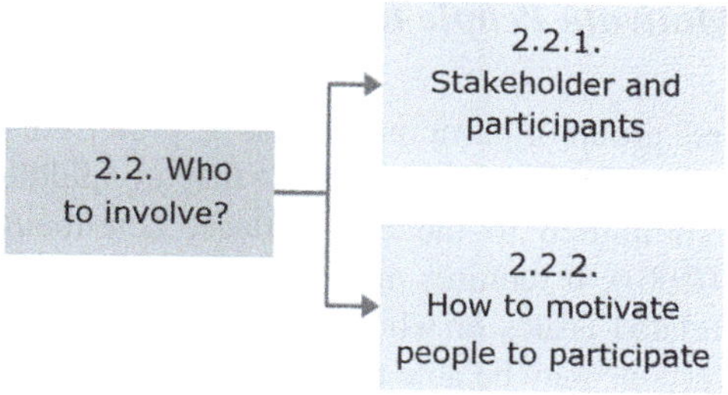

Fig. 2.6 Overview of Sect. 2.2

2.2.1 Stakeholders and Participants

Stakeholders

Local governments are often the primary actors developing and implementing climate adaptation measures to protect the local community from droughts, flooding, storms or heat stress (Mees et al. 2012). At the same time, efforts of other local actors are deemed essential to enhance climate resilience (Hegger et al. 2017). In the case of urban and rural climate resilience, stakeholders include all local actors, among which citizens, with an interest in, or who are influenced by climate impacts in their city, village, neighbourhood, or farmland. Relevant stakeholders in the context of this toolbox include *at least* local citizens and *optionally* policymakers, local NGOs, companies, and experts.

Participants

After relevant stakeholders have been identified, the next step is to decide who to invite for the participatory activity. This again relates to the goal of participation. For instance, if the goal is community building or knowledge and capacity building, a diverse, representative group of citizens could generate a realistic understanding of community needs concerning climate change. If the goal is policy development, the input of a heterogeneous group of participants could as well be valuable in early policy phases, when the problem framing of adaptation and questions of what a climate-resilient future would look like is still open for debate. In later policy phases, more expert knowledge (e.g. legal, technical) may be needed to develop adaptation measures. Organizers may then choose for a skewed representation of citizens (Uittenbroek et al. 2019).

2.2.2 *How to Motivate People to Participate?*

Even though a diverse group of participants may be preferred, participatory activities typically attract the 'usual suspects' (people that are middle-aged, interested in sustainability, and committed to the city, village, neighbourhood, or farmland (Wardekker et al. 2020b)). It requires more efforts to involve citizens without any interest in climate-related issues or who mistrust any governmental organization. Furthermore, not everyone may be able to participate in person due to time, money, or mobility constraints. Below, we provide tips to motivate also these citizens to participate.

Trust

The degree to which people trust a certain project and its organizers influences whether they want to spend time and effort on participation. Trust-building requires efforts from organizers before the participatory activity. For instance, Marschütz et al. (2020) intensively collaborated with a local cafe and neighbourhood centre. A year later, when they organized a follow-up workshop for these citizens and local policymakers to identify climate adaptation plans in the neighbourhood, only three citizens participated. The reason for this was that although citizens indeed trusted the researchers, there was still much distrust in the local municipality. Especially with regards to the policy development goal, it is therefore recommended for policymakers to actively engage with citizens too, for instance, by literally going into the neighbourhood. This personal contact helps policymakers to understand the local context, to comply with community norms and priorities, and to show their willingness (Wardekker et al. 2020b).

The Setting of the Activity

The setting of activities is an often mentioned obstacle for citizens to participate. Therefore, conscious choices regarding the frequency, duration and location of participation are to be made.

Firstly, organizers should think carefully about *how many* participatory events are needed to achieve the intended goal (see Sect. 2.1). Where the goal of community building can be met with only one participatory event, policy development goals may require several events, especially when citizens are to be engaged throughout the entire policy process (i.e. from initial agenda-setting up to the implementation of measures).

Secondly, also the *duration* of a participatory foresight activity depends on what organizers want to achieve with it. At the same time, it should be noted that the duration has implications for people's willingness to participate. Short evening sessions are generally attractive to citizens (Wardekker et al. 2020b) and can generate a representative idea about people's desires and concerns, raise awareness and provide a platform for discussion. Hence, these types of activities are particularly useful to achieve goals related to community building. However, short activities may not produce output specific enough to feed in policy planning. Longer backcasting workshops of halve a day, or even a whole day or multiple days generate more detailed material for policy development (Van Bers et al. 2016), but they also attract fewer participants, which could make the output less representative. For participatory activities aiming at policy development, this dilemma can be solved by conducting interviews (Tool 2.1), (online) surveys (Tool 2.2), or focus groups (Tool 2.3) with a larger group of citizens to get a first idea of their perspectives. This information can then be analysed with data analysis programs (Tool 2.4) and used as a basis for a longer foresight activity with fewer participants (Krauß et al. 2018; Wardekker et al. 2020b).

Finally, with regards to the *location*, experience shows that people prefer activities that are organized in their own living environment, as it leaves them in their safe space and requires less travel time. Besides practical reasons, organizing participatory activities in the neighbourhood or rural area is also a way for organizers and local governments to show their willingness to truly engage (Marschütz et al. 2020).

Note 5: *Particularly for longer sessions, it requires thinking on how to compensate participants for their attendance (e.g. with money or a gift card).*

Note 6: *Participatory foresight activities often take longer than planned, so be aware of the output that can realistically be expected.*

Relevance

Besides building trust and taking care of the physical setting for participation, another way to stimulate people to participate is by showing the relevance of the activity.

- *Examine people's motivation*

First, it is important to examine how the participatory foresight activity could be embedded in people's practices. Knowing that people have diverse priorities, how could participation be valuable for them? Experience shows that there are various reasons for participants to attend foresight activities. For instance, they may want to have a say in municipal spatial planning (Dammers et al. 2013; Uittenbroek et al. 2019). For these people, participatory activities are especially useful if linked to existing plans such as the redevelopment of a street. This way, foresight methods form a direct communicative bridge between citizens and local governments.

Participants can also be driven by social reasons: to meet other people, to network, to learn, or to have fun. These people may not be primarily interested in whether their input directly ends up in the policy process. For organizers, it is recommended to try to understand these motivations beforehand to manage expectations participants may have.

- *Climate adaptation is a collective issue*

Although the issue of climate change has received increasing attention in the public debate, it is mainly climate change *mitigation* (e.g. the energy transition) that people are aware of and act upon. For climate change *adaptation*, there has been significantly less attention (Klein et al. 2018; Mees et al. 2012). This may be because many people, especially in Western Europe, have not experienced major climate change impacts affecting their own livelihood yet. A lack of urgency obviously lowers people's motivation to participate in workshops or other activities related to climate adaptation.

Community awareness about climate adaptation can be raised with tools showing what a city, village, neighbourhood, or farmland could look like in the future under climate change. Examples of tools are posters with visual maps (Tool 2.6) showing the impacts of climate change under different scenarios (Sheppard et al. 2011), or guided city walks (Tool 2.5) showing places that are particularly vulnerable to climate change (De Voogt and Munaretto 2017). People's imagination can also be stimulated through an interactive experience (Tool 2.7) showing a glimpse of what the future could entail. Utrecht University's Urban Futures Studio built an experiential exhibition where people could literally 'walk' from the present into the future (Naafs 2018). Finally, according to Peeters (2014, pp. 98–99), also activities organised by the cultural sector (e.g. an exposition about climate change) can play an important role:

> 'The broader cultural sector can use its mobilizing potential to involve many people in an ongoing public debate concerning where we are today and where we wish to be tomorrow. This begins with continuous discussion, deliberation and reflection on one's own practice with respect to sustainable development: in-house, in dialogue with partners and the public'.

Invitation

Eventually, it is time to invite participants. Before approaching people, it is important to think about the framing of the invitation. A broad framing could attract a more diverse group of people; questions such as '*How will everyday life in your city be carried out in a much more environmentally friendly way 30 years from now?*' only require some attachment to the living environment and an interest in environmental questions, rather than specific expert knowledge (Carlsson-Kanyama et al. 2008). Also important to consider is the framing of the goal of participation. Invitations that mention participation goals related to policy development may lead prospective participants to expect that they can influence the policy process.

When the text is developed, the invitations can be distributed. A way to approach a diverse group of people is to simply approach them in their living environment (e.g. the local café, market, or square), or get in touch via local 'connectors' or hubs (e.g. community centers or similar meeting spaces, social workers, social entrepreneurs, etc.). Another way is to hang posters, distribute flyers at local stores or shopping centres or use social media channels, such as Twitter, TikTok or Instagram. A more formal and personal approach is to invite people by letter or telephone (Carlsson-Kanyama et al. 2008).

SUMMARY TOOLS **PREPARATION PHASE**

Tool 2.1 Interviews (see Appendix A for more information)

> To gather a first set of citizen perspectives as a basis for visioning activities

Tool 2.2 Surveys (see Appendix A for more information)

> To gather a first set of citizen perspectives as a basis for visioning activities

Tool 2.3 Focus groups (see Appendix A for more information)

> To gather a first set of citizen perspectives as a basis for visioning activities

Tool 2.4 Data analysis programs (see Appendix A for more information)

> To analyse citizen perspectives as a basis for visioning activities

Tool 2.5 Guided walks (see Appendix A for more information)

> To engage with the local community
> To raise awareness about climate change impacts
> To let citizens and policymakers meet in an informal way

Tool 2.6 Visual maps (see Appendix A for more information)

> To raise awareness about how climate change may impact the local context

Tool 2.7 Future experiences (see Appendix A for more information)

> To raise awareness about climate change through experiential imagination

2.3 Which Foresight Methods to Use?

The term *foresight* refers to a wide range of methods for exploring and engaging with the future (Hebinck et al. 2018). In this section, we discuss some of these foresight methods and their potential role in building a local climate-resilient future. The choice of which methods to use depends on the way you want participants to interact with the future.

Do you want participants to:

- … *explore* what the future could bring? **Exploratory scenarios** (Sect. 2.3.1) describe how futures might evolve under several drivers of change such as temperature rise, extreme weather events, or economic growth (see Fig. 2.7, left)
- … *shape* a more desired future? **Visioning** methods (Sect. 2.3.2) invite participants to envision a climate-resilient living environment (see Fig. 2.7, right: the yellow circles are visions)

- ... *formulate concrete actions* to achieve a climate-resilient living environment in the future? Then **pathways of change** (Sect. 2.3.3) can be developed by participants (see Fig. 2.7, right: the dotted lines are pathways).

In the next sections, we explain these three foresight methods and discuss how they lend themselves to be used in participatory activities with different participation goals (see Fig. 2.8 for a visualized overview of Sect. 2.3). In practice, organizers often use a combination of foresight methods (Hebinck et al. 2013, 2018; Raudsepp-Hearne et al. 2019; Wardekker et al. 2020b). We discuss possible combinations of foresight methods in Sect. 2.3.4.

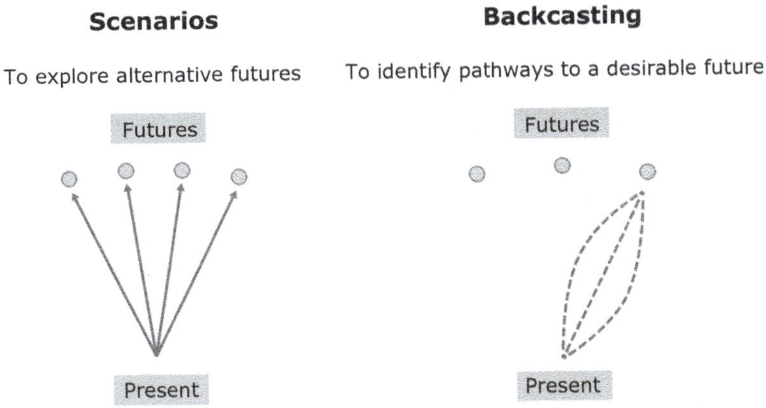

Fig. 2.7 Different perceptions of the future. (Adapted from Van Bers et al. 2016)

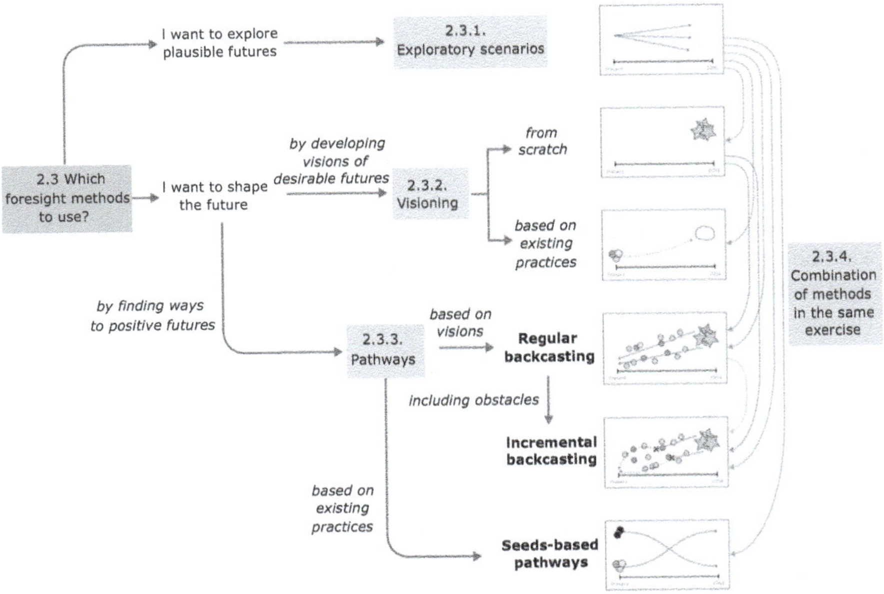

Fig. 2.8 Overview of Sect. 2.3

2.3.1 *Exploratory Scenarios*

2.3.1.1 What Are Exploratory Scenarios?

Scenarios have an explorative character: they describe a range of futures that *may* happen (see Fig. 2.9). Exploratory scenarios are often used based on the idea that, in complex systems and under high uncertainty, it is impossible to predict the most likely future. Instead, with multiple plausible (socioeconomic or climate) scenarios 'what if' questions can be explored (Wiebe et al. 2018), such as 'what would these different situations mean for our attempts to implement climate adaptation measures?', or 'what would low/high precipitation levels mean for this farmland in the future?'.

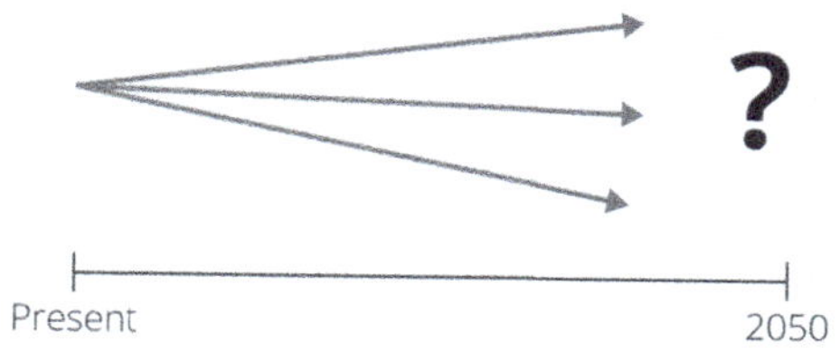

Fig. 2.9 Exploratory scenarios of plausible futures

2.3.1.2 Why Use Exploratory Scenarios?

- **Participation goal 1: Policy development.** Exploratory scenarios can assess the feasibility of visions and pathways.

In the context of this toolbox, the ultimate goal of organizing participatory foresight activities is to contribute to local climate resilience. However, exploratory scenarios are based on dominant trends and projections of 'what could happen' are not necessarily climate-resilient (Carlsson-Kanyama et al. 2008). Still, scenarios play an important role in building a better future. For instance, policymakers can test adaptation measures under scenario A and under scenario B to ensure the robustness of climate adaptation plans (Dammers et al. 2013; Vervoort and Mangnus 2018; Wardekker et al. 2020b; Wiebe et al. 2018).

In the context of this toolbox, exploratory scenarios are thus particularly useful in combination with other foresight methods, such as visioning and backcasting.

- **Participation goal 2: Community building.** Exploratory scenarios build awareness about possible climate change impacts.

Participatory activities with exploratory scenarios can let participants experience the 'realness' of changes in climate, socioeconomic and technological trends and potential effects on their daily life (Carlsson-Kanyama et al. 2008; Henstra 2012). For instance, in urban areas, the vulnerability of specific streets can be shown under different climate scenarios (Wardekker et al. 2020b). In rural areas, seasonal forecasts can generate insights on possible impacts on yield, production, and livestock diseases, based on which adaptation strategies can be developed (CARE International 2018; Wiebe et al. 2018). Information about climate vulnerabilities provided by exploratory scenarios can be seen as a climate service that raises awareness about the need for adaptation. Awareness can again translate into more climate-adaptive behaviour by citizens, for instance, if they remove pavement in their gardens, install green roofs, or if they consider alternative agriculture practices (CARE International 2018; Wardekker et al. 2020b).

- **Participation goal 3: Knowledge & capacity building.** Also: Exploratory scenarios build awareness about possible climate change impacts.

2.3.2 *Visions of a Desired Future*

2.3.2.1 What Is Visioning?

Visioning as a foresight method is useful if exploratory scenarios project primarily undesired futures. In the context of this toolbox, visions encompass perceptions of what a climate-resilient living environment (e.g. a city, village, neighbourhood, or farmland) would ideally look like according to participants. Visions can be broad descriptions or they include specific measures or targets, based on new ideas (see Fig. 2.10, left) or inspired by existing local initiatives (also called 'seeds') (see Fig. 2.10, right).

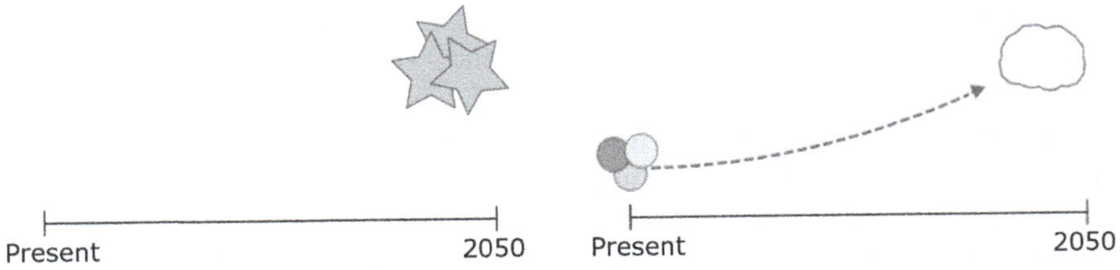

Fig. 2.10 Visions of the desired future based on new ideas (left) or deduced from present-day initiatives (right)

2.3.2.2 Why Visioning?

- **Participation goal 1: Policy development.** Visioning activities can let policymakers consider citizens' wishes and desires in the process of developing climate adaptation plans.

In the context of spatial redevelopment, policymakers can ask citizens to envision their desired living environment. For instance, questions such as '*What will this street look like in the future when we focus on climate-resilience?*' or '*What will this neighbourhood look like in the future when we focus on sustainability?*' can bring innovative insights in early policymaking stages (Hebinck et al. 2013; Wiek and Iwaniec 2014). With their local knowledge, citizens can expose blind spots or opportunities for climate adaptation that may otherwise be missed. Discussions with citizens also reveal community norms and values that need to be respected. This way, participation helps local governments develop climate action plans that citizens find meaningful and worth bringing around (Wardekker et al. 2020b). In general, the quality of visions for adaptation planning increases when they meet the following quality criteria (Wiek and Iwaniec 2014, p. 500):

- Visionary; reflecting normative elements of desirability, out-of-the-box thinking and holistic perspectives
- Sustainable; reflecting elements of transformations toward sustainability
- Systemic; reflecting elements of holistic and complex thinking
- Coherent; reflecting logical/compatible objectives
- Plausible; reflecting empirical evidence. Tip: Use existing sustainable initiatives (seeds) (Ramos et al. 2019; Raudsepp-Hearne et al. 2019)
- Tangible; reflecting clearly described objectives
- Relevant; reflecting objectives that fit the local context
- Nuanced; reflecting detailed desires
- Motivational; reflecting inspiration for change
- Sharede; reflecting collective agreement by key stakeholders. Tip: Involve policymakers in co-designing visions (Schubert 2014)

- **Participation goal 2: Community building.** Visioning activities can raise awareness, stimulate citizen action and build public support for new climate action plans.

The visioning *process* itself contributes to community building: it raises awareness and stimulates social learning; both essential for gaining public support for the implementation of plans (Ramos et al. 2019; Wiek and Iwaniec 2014). This way, visioning activities indirectly contribute to the policy development goal. Besides public support, visioning activities also provide 'room for imagination', which according to Peeters (2014, p. 99) is essential:

> 'Without imagination, inspiration and creativity, we are unable to create future visions for a new society. 'From which view of the world, humanity and life do we wish to develop a resilient society?' Imagination, via a shared horizon, gives meaning to human activities, and allows trying out new ideas, lifestyles and perspectives.'

Hence, visioning activities could give people a sense of hope, meaning, and inspiration, which could translate into citizen initiatives, networks, collective action or other forms of public mobilization. Further, people are more likely to perceive themselves as 'owner' of the problem and solution if they are actively involved in formulating local issues and shaping future desires (see Fig. 2.11) (Uittenbroek et al. 2019).

Fig. 2.11 Participant in a visioning activity. (©Studio Lakmoes)

2.3.3 Pathways to the Desired Future

2.3.3.1 What Are Pathways?

Having developed a vision of a climate-resilient living environment, the next step is to identify *ways to get there* in terms of actual solutions, actions, or measures (Van Bers et al. 2016; Wardekker et al. 2020b). These normative pathways are a counterpart to exploratory scenarios that sketch plausible futures which are out of local control (Van Notten et al. 2003).

There are several ways to develop pathways. In this toolbox, we discuss backcasting, incremental backcasting, and seeds-based pathways.

Backcasting pathways start from the desired future vision and form a sequence of short, medium and long-term actions back to the present (Van Bers et al. 2016; Wardekker et al. 2020b). When categorizing these actions in themes and placing them on a timeline, thematic pathways can be identified (see Fig. 2.12).

Fig. 2.12 A backcasting pathway is a sequence of actions that connects the future with the present

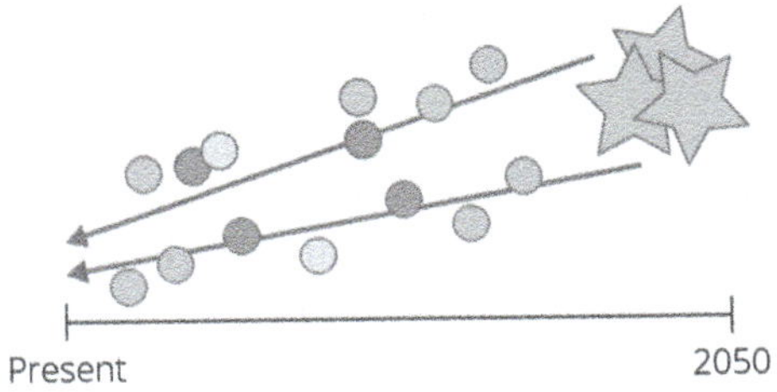

Backcasting activities thus form more or less straight paths from the future to the present. However, researcher Wardekker argues that in reality, numerous constraints can happen along the way (Ballard 2019):

> 'The future rarely unfolds linearly. Along the way, things can go wrong, or there are instances where you can make use of new opportunities that present themselves. And there can always be surprises.'

Such disturbances, positive or negative, are called 'hinge-points'. The term relates to the concept of 'trigger', which is commonly used in the field of policy planning (Haasnoot et al. 2013). It specifies the condition in which a particular pathway of actions is no longer adequate for achieving the desired goal, or when new opportunities arise that accelerate the process. Hinge-points can be internal and controllable (e.g. the construction of a new sewage system), but also external and uncontrollable (e.g. political ideology, or extreme sea-level rise). Either way, these unexpected events could steer backcasting pathways in a different direction (see Fig. 2.13). To account for this, Wardekker et al. (2020a, b) introduced a novel way of developing pathways that anticipate sudden disturbances, also called **incremental backcasting pathways.**

Backcasting pathways and incremental pathways start from a future vision and go back to the present, but there is also a way of developing pathways from the present to the future. It builds on positive, local initiatives that already exist and have proven to be successful in other contexts – also referred to as 'seeds' (Bennett et al. 2016; Raudsepp-Hearne et al. 2019). **Seeds-based pathways** start in the present and explore what is needed for local initiatives on climate adaptation to successfully grow in the future (see Fig. 2.14). The interesting aspect of this approach is that although the pathways are based on ongoing activities and practices in the present, they are potentially transformative in the future (Bahadur and Tanner 2014; Vervoort and Mangnus 2018).

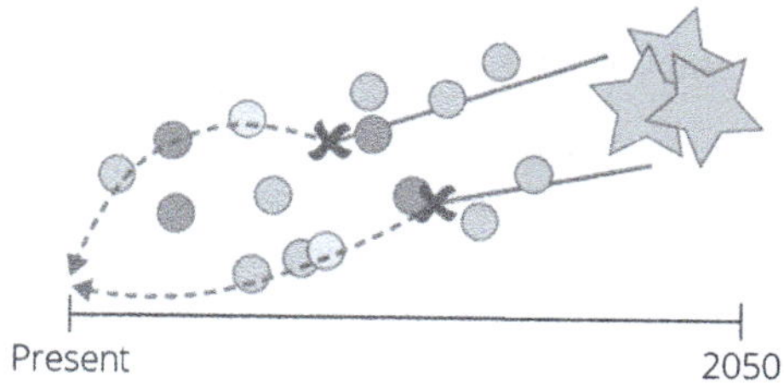

Fig. 2.13 Incremental backcasting pathways are backcasting pathways that anticipate sudden disturbances or chances

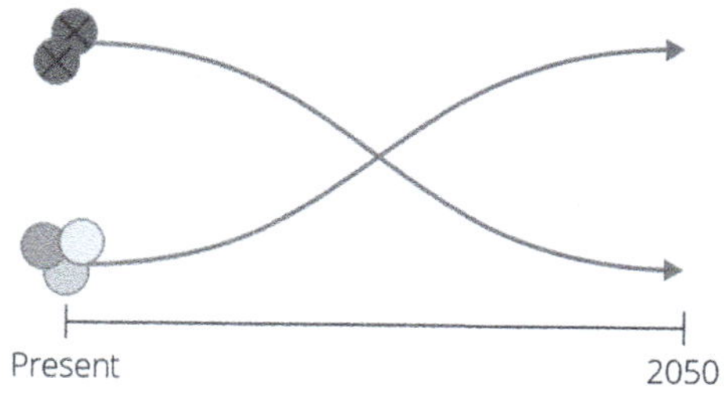

Fig. 2.14 Seeds-based pathways: positive and negative elements of the present that should grow and decline, respectively, for a more climate-resilient future

2.3.3.2 Why Backcasting?

- **Participation goal 1: Policy development.** Backcasting pathways developed by citizens contain diverse and innovative ideas for climate adaptation plans.

In backcasting activities, participants formulate actions that they deem relevant for achieving a climate-resilient living environment. Hence, they can provide additions to adaptation plans that scientists and policymakers may not be able to capture (Dammers et al. 2013). Policymakers can thus organize a backcasting activity with citizens if they seek inspiration for more innovative, out-of-the-box climate strategies that also better fit the local context (Uittenbroek et al. 2019). This is especially useful in early policy phases. Finally, by contributing to the design of policy, citizens can become more supportive of the implementation.

- **Participation goal 2: Community building.** Backcasting pathways let citizens experience how their desired climate-resilient future can become reality.

Similar to exploratory scenarios and visioning activities, backcasting activities can raise awareness and stimulate learning about climate change. The general idea is that backcasting activities where citizens develop their pathways bring a higher buy-in than if scenarios are only presented to them (Carlsson-Kanyama et al. 2008). Having thought about what it takes to achieve a climate-resilient living environment, people are likely to make more climate-adaptive choices in and round their own house as well. Ideas and inspiration gained during the backcasting activity can even be the start of new collaborations and partnerships between citizens, local governments and businesses.

2.3.3.3 Why Incremental Backcasting?

- **Participation goal 1: Policy development.** Incremental pathways expose potential obstacles to climate adaptation plans.

Whereas backcasting activities generate linear sequences of actions from the future to the present, incremental pathways consider the likelihood of sudden hinge-points. Participants then come up with alternative action plans to still be able to reach the desired future. Incremental pathways are particularly useful for policy-related goals of participation; they help policymakers anticipate various disturbances with robust and reflexive actions that prevent the climate action plan from crashing (Haasnoot et al. 2020; Wardekker et al. 2020a, b). As such, incremental pathways explicitly link foresight with policy planning.

- **Participation goal 3: Knowledge and capacity building.** Incremental pathways help identify climate service needs to overcome obstacles.

Incremental backcasting activities are also useful to identify climate service needs that community members need to overcome a disturbance. An example is specific information to better communicate the issue of climate change at moments when it is particularly needed, such as during disasters (Wardekker et al. 2020a, b). Particularly knowledge producers, such as climate experts and scientists, may want to know climate service needs to ensure the relevance and usefulness of information (Bremer et al. 2019). Hence, rather than just providing climate information, they can look at climate services through the lens of what is actually needed. Or, as researcher Arjan Wardekker formulated: 'It is not about dryly communicating about their neighborhood and how it's affected by weather, climate and water. It's about moving from matters of fact to matters of concern' (Ballard 2019).

2.3.3.4 Why Seeds-Based Pathways?

- **Participation goal 1: Policy development.** Seeds-based pathways show evidence-based routes to climate-resilient futures.

While backcasting has proven to be useful in many ways, a drawback experienced by participants is the difficulty to come up with concrete actions. Some also struggle with finding a balance between a truly different, more sustainable living environment that is also realistic. To avoid visions that are either 'stuck in the present' or almost 'utopian' or 'science fiction', organizers can let participants develop pathways based on seeds: sustainability initiatives that have proven to be successful elsewhere (Bennett et al. 2016). Seeds-based pathways are potentially transformative and plausible and nuanced at the same time, simply because they are grounded in present-day realities. Such methods can make policymakers aware of conditions that enable good initiatives to grow and scale-up in their own local context (Raudsepp-Hearne et al. 2019).

- **Participation goal 2: Community building.** Seeds-based pathways let citizens experience the possibility of transforming their living environment.

Seeds-based pathways do not necessarily need to generate output relevant to policy (Raudsepp-Hearne et al. 2019). The value of seeds-based pathways for the local community is that they can let people realize that completely different (more positive) futures are feasible and that every small step counts (Pereira et al. 2018).

2.3.4 Combinations of Methods in the Same Activity

Participatory foresight activities generally produce more detailed output if various methods are used in a complementary way. Figure 2.15 gives an overview of relevant combinations of foresight methods. The usefulness of each combination is explained in Table 2.1.

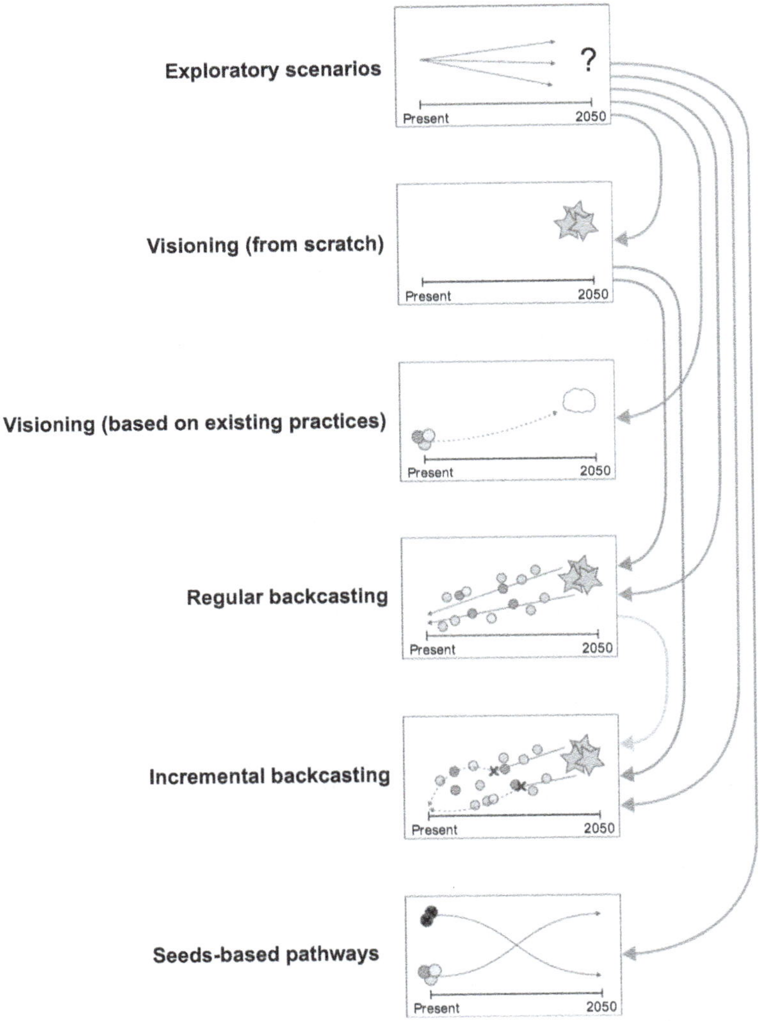

Fig. 2.15 Overview of which foresight methods can complement others in a participatory activity

Table 2.1 The relevance of combining foresight methods in a participatory activity

	Exploratory scenarios	Visioning	Backcasting	Incremental backcasting	Seeds-based pathways
Exploratory scenarios activity		x	x	x	x
Visioning activity	To develop visions of a desired future while taking into account possible implications of different scenarios		To formulate concrete actions to reach visions of a desired future	To anticipate hinge-points that may disturb or accelerate backcasting pathways on their way to reach visions of a desired future	x
Backcasting activity	To develop concrete actions to reach visions of a desired future while taking into account possible implications of different scenarios	Visions of a desired future are the starting point from which to develop backcasting pathways		To anticipate hinge-points that may disturb or accelerate backcasting pathways on their way to reach visions of a desired future	x
Incremental backcasting activity	To use elements of different scenarios as inspiration source for formulating hinge-points	Visions of a desired future are the starting point from which to develop backcasting pathways and then incremental backcasting pathways	Backcasting pathways are the basis of incremental backcasting pathways		x
Seeds-based pathways activity	To develop seeds-based pathways while taking into account possible implications of different scenarios	x	x	x	

Appendix A

Additional Information About Tools for Preparing Participatory Foresight Activities

Tool 2.1: Interviews

- LEVEL ORGANIZER: ■■□
- LEVEL PARTICIPANT: ■□□
- DURATION: 30–60 mins

What? In preparation for participatory foresight activities, organizers can conduct interviews to get a better understanding of the local context. The *process* of conducting interviews can contribute to trust building and it creates an opening when participants need to be recruited for the actual participatory foresight activity.

When? In the preparation phase.

How? People can be asked about their experiences with climate change impacts such as heavy rainfall or heat waves, and how this has influenced their daily life (Marschütz et al. 2020). They can also be asked about climate-related fears and desires for the future, which can be used as a basis for visioning activities (Wardekker et al. 2020b). This way, a large group of stakeholders can be represented without the need to physically attend the participatory foresight activity.

→ **For example questions, see Marschütz et al. (2020, pp. 161–162)**

Tool 2.2: Surveys

- LEVEL ORGANIZER: ■□□
- LEVEL PARTICIPANT: ■□□
- DURATION: 15–30 mins

What? In preparation for participatory foresight activities, organizers can conduct surveys to get an understanding of the local context. Surveys do not necessarily require personal contact; they can also be held online to reach a larger group of people. However, a lack of personal contact also means less trust building.

When? In the preparation phase.

How? People can be asked about their experiences with climate change impacts such as heavy rainfall or heat waves, and how this has influenced their daily life (Marschütz et al. 2020). They can also be asked about climate-related fears and desires for the future, which can be used as a basis for visioning activities (Wardekker et al. 2020b). This way, a large group of stakeholders can be represented without the need to physically attend the participatory foresight activity.

→ **For example questions, see Marschütz et al. (2020, pp. 161–162)**

Tool 2.3: Focus Groups

- LEVEL ORGANIZER: ■■■
- LEVEL PARTICIPANT: ■□□
- DURATION: 1–2 h

What? In preparation for participatory foresight activities, organizers can conduct a focus group (i.e. group interviews) to get an understanding of the local context. They lend themselves for elaborated discussions among stakeholders and as such they can contribute to networking and trust-building between organizers, stakeholders and policymakers. Focus groups usually include a maximum of 10 participants.

When? In the preparation phase.

How? People can be asked about their experiences with climate change impacts such as heavy rainfall or heat waves, and how this has influenced their daily life (Marschütz et al. 2020). They can also be asked about climate-related fears and desires for the future, which can be used as a basis for visioning activities (Wardekker et al. 2020b). This way, a large group of stakeholders can be represented without the need to physically attend the participatory foresight activity.

→ **For example questions, see Marschütz et al. (2020, pp. 161–162)**

Tool 2.4: Data Analysis Programs

- LEVEL ORGANIZER: ■■■
- LEVEL PARTICIPANT: -
- DURATION: 2+ h

What? Interviews, surveys and focus groups generate a large set of data in terms of citizen perspectives. These narratives can be analyzed with programs to use a basis for participatory foresight activities (Krauß et al. 2018).

When? In the preparation phase.

How? There are various data analysis programs that can be used to code, analyze or cluster citizen perspectives. One example is wordcloud, a program to visualize word frequency diagrams.

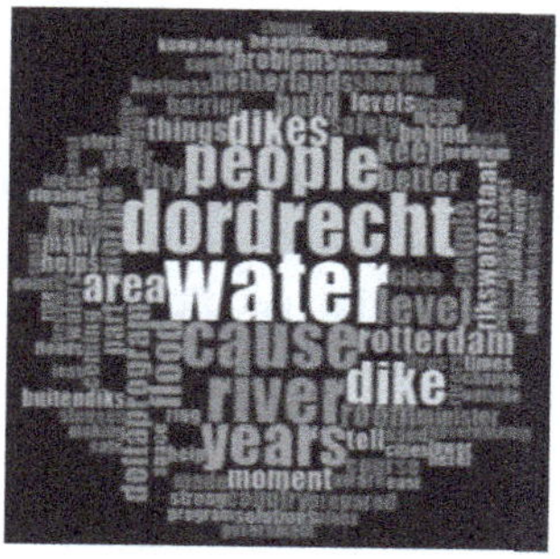

Example of a wordcloud of narrative interviews. From Krauß et al. (2018).

Tool 2.5: Guided Walks

- LEVEL ORGANIZER: ■■■
- LEVEL PARTICIPANT: ■□□
- DURATION: 1–2 h

What? Organizing guided walks with citizens and policymakers in the neighbourhood or city is an (inter)active way to share knowledge about climate change impacts, listen to people's experiences and generate a sense of urgency among the local community.

When? In the preparation phase.

How? Experts can show specific places in the neighbourhood that are vulnerable to climate change impacts. Optionally, smartphones, tablets with Augmented Reality or visual maps can be used to show on the spot what extreme weather events could mean for the area as a way to stimulate participants' imagination during the walk. Also citizens can organize a walk, e.g. for fellow citizens and policymakers, to show what they care about and express their desires and fear in relation to climate change (De Voogt and Munaretto 2017). They can also bring photos to show the effects of previously experienced climate change impacts. Furthermore, walks allow citizens and local policymakers to meet in an informal way. Guided walks can also be organized for elementary schools (children often share their learning experiences with their parents!).

Tool 2.6: Visual Maps

- LEVEL ORGANIZER: ■■■
- LEVEL PARTICIPANT: ■□□
- DURATION: 30–60 mins

What? Visual GIS maps (e.g. 2D, 3D, fly-over) of the local living environment under different (climate) scenarios can be used as a tool to imagine possible future situations. These maps could contain any information about climate change and how it affects the local area, from flooding to extreme heat. Visual maps make use of people's emotional connection to the place and as such bring climate change to life (Sheppard et al. 2011).

When? In the preparation phase.

How? Visual maps showing possible climate change impacts in highly different situations (e.g. heat stress visualized for a green street full of trees and for a paved street without trees) can be presented to people on posters in the neighbourhood or in local newspapers to raise awareness about the importance of taking climate adaptation measures.

A visual map showing a neighborhood in Dordrecht under the impact of heat stress. From Wardekker et al. (2020b).

Tool 2.7: Future Experiences

- LEVEL ORGANIZER: ■■■
- LEVEL PARTICIPANT: ■□□
- DURATION: 1–2 h

What? Interactive experiences can let people experience different futures with all their senses. Such simulations of the future trigger out-of-the-box thinking and provide a space for lively discussions, learning, awareness raising and networking.

When? In the preparation phase.

How? Examples of future experiences are model homes or prototypes that literally provide a glimpse of what could be mainstream in the future. Another example is a project by Utrecht University ('Neighbourhoods for the Future') that shows the innovation potential of neighbourhoods (Naafs 2018). Hundreds of neighbourhoods innovation around mobility, energy, water, inclusivity and circularity were listed, of which some have been presented at an experimental exhibition called 'Places of Hope'. The Urban Futures Studio of Utrecht University, together with urban designers (Non-Fiction and The Cloud Collective), initiated a novel technique of letting people experience the future physically and emotionally by literally 'walking' into it.

→ **For more information about the exhibition 'Placed of Hope', see** https://vimeo.com/333718626

References

Arnstein SR (1969) A ladder of citizen participation. J Am Inst Plann 35(4):216–224. https://doi.org/10.1080/01944366908977225

Bahadur A, Tanner T (2014) Transformational resilience thinking: putting people, power and politics at the heart of urban climate resilience. Environ Urban 26(1):200–214. https://doi.org/10.1177/0956247814522154

Ballard C (2019) Engaging citizens for a climate-resilient Dordrecht. https://www.uu.nl/en/news/engaging-citizens-for-a-climate-resilient-dordrecht

Bennett EM, Solan M, Biggs R, McPhearson T, Norström AV, Olsson P, Pereira L, Peterson GD, Raudsepp-Hearne C, Biermann F, Carpenter SR, Ellis E, Hichert T, Galaz V, Lahsen M, Milkoreit M, Martin López B, Nicholas KA, Preiser R et al (2016) Bright spots: seeds of a good Anthropocene. Front Ecol Environ 14(8):441–448. https://doi.org/10.1002/fee.1309

Bremer S, Wardekker A, Dessai S, Sobolowski S, Slaattelid R, Van der Sluis J (2019) Toward a multi-faceted conception of co-production of climate services. Clim Serv 13:42–50. https://doi.org/10.1016/j.cliser.2019.01.003

Bremer S, Wardekker A, Jensen ES, van der Sluijs JP (2021) Quality assessment in co-developing climate Services in Norway and the Netherlands. Front Clim 3:627665. https://doi.org/10.3389/fclim.2021.627665

CARE International (2018) Practical guide to participatory scenario planning: seasonal climate information for resilient decision-making. CARE International, Geneva

Carlsson-Kanyama A, Dreborg KH, Moll HC, Padovan D (2008) Participative backcasting: A tool for involving stakeholders in local sustainability planning. Futures 40:34–46. https://doi.org/10.1016/j.futures.2007.06.001

Cash DW, Clark WC, Alcock F, Dickson NM, Eckley N, Guston DH, Jäger J, Mitchell RB (2003) Knowledge systems for sustainable development. PNAS 100(14):8086–8091. https://doi.org/10.1073/pnas.1231332100

Crawford MM (2019) A comprehensive scenario intervention typology. Technol Forecast Soc Change 149:119748

Da Costa O, Warnke P, Cagnin C, Scapolo F (2008) The impact of foresight on policy-making: insights from the FORLEARN mutual learning process. Tech Anal Strat Manag 20(3):369–387. https://doi.org/10.1080/09537320802000146

Dammers E, Van 't Klooster S, De Wit B, Hilderink H, Petersen A, Tuinstra W (2013) Scenario's maken voor milieu, natuur en ruimte: een handreiking. PBL, Den Haag

De Voogt DL, Munaretto S (2017) Participatory mechanisms report for the Dutch case study in Itteren and Borgharen. Planning, implementation and evaluation of pilot actions: CAPFLO project report – final report https://drive.google.com/file/d/0B8hFBvkmQSTRTzNSQ3UtUTcyalk/view

Dinesh D, Hegger DLT, Vervoort JM, Driessen PPJ (2021) A changing climate for knowledge generation in agriculture: lessons to institutionalize science-policy engagement. Front Clim 3:615463. https://doi.org/10.3389/fclim.2021.615463

Haasnoot M, Kwakkel JH, Walker WE, Ter Maat J (2013) Dynamic adaptive policy pathways: A method for crafting robust decisions for a deeply uncertain world. Glob Environ Chang 23:485–498. https://doi.org/10.1016/j.gloenvcha.2012.12.006

Haasnoot M, Biesbroek R, Lawrence J, Muccione V, Lempert R, Glavovic B (2020) Defining the solution space to accelerate climate change adaptation. Reg Environ Chang 20(37):1–5. https://doi.org/10.1007/s10113-020-01623-8

Hebinck A, Villarreal G, Oostindie H, Hebinck P, Zwart TA, Vervoort J, Rutting L, De Vrieze A (2013) Urban agriculture policy-making: Proeftuin040 – TRANSMANGO scenario workshop report, the Netherlands

Hebinck A, Vervoort JM, Hebinck P, Rutting L, Galli F (2018) Imagining transformative futures: participatory foresight for food systems. Ecol Soc 23(2):16

Hegger DLT, Mees HLP, Driessen PPJ, Runhaar HAC (2017) The roles of residents in climate adaptation: A systematic review in the case of the Netherlands. Environ Policy Gov 27:336–350. https://doi.org/10.1002/eet.1766

Henstra D (2012) Toward the climate-resilient city: extreme weather and urban climate adaptation policies in two Canadian provinces. J Compar Policy Anal Res Pract 14(2):175–194. https://doi.org/10.1080/13876988.2012.665215

Klein J, Araos M, Karimo A, Heikkinen M, Ylä-Anttila T, Juhola S (2018) The role of the private sector and citizens in urban climate change adaptation: evidence from a global assessment of large cities. Glob Environ Chang 53:127–136. https://doi.org/10.1016/j.gloenvcha.2018.09.012

Krauß W, Bremer S, Wardekker A, Marschütz B, Baztan J, Da Cunha C (2018) Initial mapping of narratives of change. CoCliServ, Guyancourt, France.
Lorenz S, Dessai S, Forster PM, Paavola J (2017) Adaptation planning and the use of climate change projections in local government in England and Germany. Reg Environ Chang 17:425–435. https://doi.org/10.1007/s10113-016-1030-3
Marschütz B, Bremer S, Runhaar H, Hegger D, Mees H, Vervoort J, Wardekker A (2020) Local narratives of change as an entry point for building urban climate resilience. Clim Risk Manag 28:100223. https://doi.org/10.1016/j.crm.2020.100223
Mees HLP, Driessen PPJ, Runhaar HAC (2012) Exploring the scope of public and private responsibilities for climate adaptation. J Environ Policy Plan 14(3):305–330. https://doi.org/10.1080/1523908X.2012.707407
Mees HLP, Driessen PPJ, Runhaar HAC, Stamatelos J (2013) Who governs climate adaptation? Getting green roofs for stormwater retention off the ground. J Environ Plan Manag 56(6):802–825. https://doi.org/10.1080/09640568.2012.706600
Naafs S (2018) Spetterende slotmanifestatie Places of Hope: Zin in de toekomst! https://placesofhope.nl/nieuws/als-het-zo-kan-dan-wil-ik-het-wel/
Peeters J (2014) Imagination, Experience and Meaning as Quality of Life. In: Cools G and Gielen P (eds.) The Ethics of Art: Ecological Turns in the Performing Arts (pp.89-107). Valiz, Amsterdam, The Netherlands.
Pereira LM, Hichert T, Hamann M, Preiser R, Biggs R (2018) Using futures methods to create transformative spaces: visions of a good Anthropocene in southern Africa. Ecol Soc 23(1):19. https://doi.org/10.5751/ES-09907-230119
Ramos J, Sweeney JA, Peach K, Smith L (2019) Our futures: by the people, for the people. https://media.nesta.org.uk/documents/Our_futures_by_the_people_for_the_people_HrqsGPo.pdf
Raudsepp-Hearne C, Peterson GD, Bennett EM, Biggs R, Norström AV, Pereira L, Vervoort J, Iwaniec DM, McPhearson T, Olsson P, Hichert T, Falardeau M, Jiménez Aceituno A (2019) Seeds of good anthropocenes: developing sustainability scenarios for Northern Europe. Sustain Sci. https://doi.org/10.1007/s11625-019-00714-8
Schubert C (2014) Helping Honduras build a more robust climate adaptation strategy for the agriculture sector. https://ccafs.cgiar.org/news/helping-honduras-build-more-robust-climate-adaptation-strategy-agriculture-sector
Sheppard SRJ, Shaw A, Flanders D, Burch S, Wiek A, Carmichael J, Robinson J, Cohen S (2011) Future visioning of local climate change: A framework for community engagement and planning with scenarios and visualisation. Futures 43:400–412. https://doi.org/10.1016/j.futures.2011.01.009
Tosun J, Schoenefeld JJ (2017) Collective climate action and networked climate governance. WIREs Clim Change 8. https://doi.org/10.1002/wcc.440
Uittenbroek CJ, Mees HLP, Hegger DLT, Driessen PPJ (2019) The design of public participation: who participates, when and how? Insights in climate adaptation planning from the Netherlands. J Environ Plan Manag 62(14):2529–2547. https://doi.org/10.1080/09640568.2019.1569503
Van Bers C, Bakkes J, Hordijk L (2016) TIAS report series. Building bridges from the present to desired futures. Evaluating approaches for visioning and backcasting based on a workshop held at Central European University, Budapest, Hungary 21–22 March 2011 (Issue 1)
Van den Ende MA, Mees, HLP, Hegger DLT, Driessen PJ (2022, forthcoming) Mechanisms influencing mainstreaming of adaptation in spatial development: case studies in three Dutch municipalities. J Environ Plann Manag
Van der Hel S, Biermann F (2017) The authority of science in sustainability governance: A structured comparison of six science institutions engaged with the sustainable development goals. Environ Sci Policy 77:211–220. https://doi.org/10.1016/j.envsci.2017.03.008
Van Notten PWF, Rotmans J, Van Asselt MBA, Rothman DS (2003) An updated scenario typology. Futures 35:423–443. https://doi.org/10.1016/S0016-3287(02)00090-3
Vaz F, Prendeville S (2019) Design as an agent for public policy innovation. Academy for Design Innovation Management. https://doi.org/10.33114/adim.2019.06.231

Vervoort JM, Mangnus AAC (2018) The roles of new foresight methods in urban sustainability transformations: A conceptual framework and research agenda. Urban Futures Studio, Utrecht, The Netherlands.

Wardekker A, Bremer S, Krauß W, Da Cunha C, Paula Farias Rocha A, Baztan J, Jaffrès L, Breton F, Runhaar H, Vervoort J, Van der Sluijs J, Wildschut D, Vanderlinden J-P (2020a) Protocol for designing incremental scenarios. CoCliServ, Guyancourt, France.

Wardekker A, Van den Ende M, Marschutz B, Pijnappels M, Hofland S, Bremer S, Blanchard A, Iversen L, Van der Sluis J, Krauß W, Rocha A, Da Cunha C, Baztan J, Jaffrès L (2020b) Incremental scenario case studies. CoCliServ, Guyancourt, France.

Wiebe K, Zurek M, Lord S, Brzezina N, Gabrielyan G, Libertini J, Loch A, Thapa-Parajuli R, Vervoort J, Westhoek H (2018) Scenario development and foresight analysis: exploring options to inform choices. Annu Rev Environ Resour 43:545–570. https://doi.org/10.1146/annurev-environ-102017-030109

Wiek A, Iwaniec D (2014) Quality criteria for visions and visioning in sustainability science. Sustain Sci 9:497–512. https://doi.org/10.1007/s11625-013-0208-6

Chapter 3
Performing Participatory Foresight Methods

Abstract After having prepared the participatory activity using the guiding questions of *why citizen participation?, who to involve?,* and *which foresight methods to use?* (see Chap. 2), it is time to gain an understanding of *how to use foresight methods* in practice. This chapter gives a step-by-step explanation about how to use exploratory scenarios (Sect. 3.1), followed by visioning (Sect. 3.2), backcasting (Sect. 3.3), incremental backcasting (Sect. 3.4) and seeds-based pathways (Sect. 3.5) (see Fig. 3.1).

Keyword Toolbox · Citizen participation · Futures · Foresight methods · Scenarios · Visioning · Pathways · Resilience · Climate adaptation · Urban · Rural · Sustainability

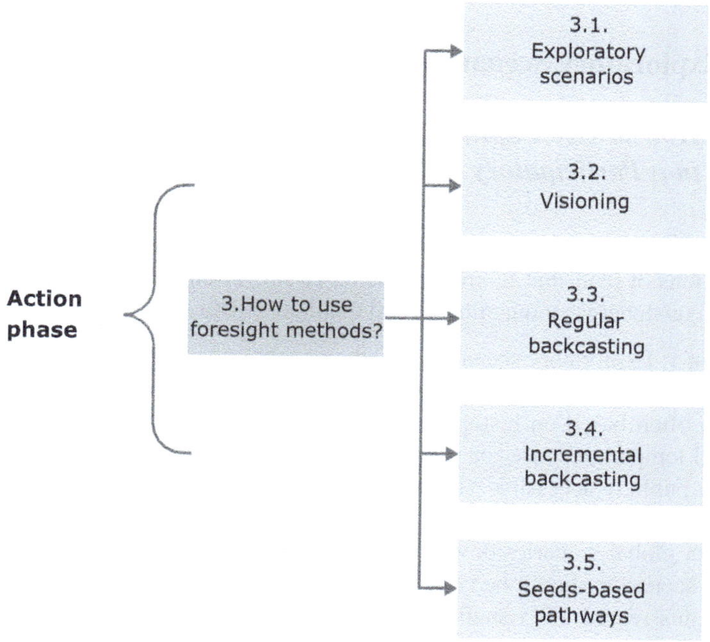

Fig. 3.1 Overview of this chapter

M. A. van den Ende et al., *Towards a Climate-Resilient Future Together*, SpringerBriefs in Environmental Science,
https://doi.org/10.1007/978-3-031-07682-4_3

Before choosing a foresight method, it is recommended to first determine the scope of the foresight activity in terms of the time horizon and the geographical area (Dammers et al. 2013; Phdungsilp 2011; Van Notten et al. 2003).

Time Horizon
Foresight activities in relation to climate change typically require a long time frame (25–50 years). However, foresight activities that focus on the far away future only may be of less interest for policymakers, hence they risk the chance of becoming a theoretical activity rather than a basis for policy (Wardekker et al. 2020a, b). For most citizens, a time horizon of maximum 40 years is still relevant as it concerns their life and that of their children (Robinson 2003). In addition to a long time frame, it is recommended to use a short/medium time frame (0–25 years) as well to zoom in on concrete measures (Van Notten et al. 2003).

Geographical Scope
In the context of this toolbox, thinking about the future requires a focus on the local level (e.g. a city, neighbourhood, river basin, or farm) as this is most relevant for citizens and policymakers (CARE International 2018; Crawford 2019; Van Notten et al. 2003). At the same time, global and national scenarios can play an important role to assess local visions and pathways on their feasibility (Carlsson-Kanyama et al. 2008; Van Notten et al. 2003).

3.1 Exploratory Scenarios

3.1.1 How to Use Exploratory Scenarios in a Participatory Activity?

- **Step 1: Explore Plausible Futures**

Explorations of plausible futures can be based on various sources. In this toolbox, we distinguish between quantitative and qualitative data.

- Option 1: Exploratory scenarios based on quantitative data

Climate scenarios developed by renowned national and international research institutes are often based on historic trends that are extrapolated into the future (e.g. projected temperature rise, or precipitation levels) (Lorenz et al. 2017). Many of them are publicly accessible via online data bases (see e.g. http://www.klimaateffectatlas.nl/en/ for Dutch climate scenarios and IPCC (2018) and UN Environment (2019) for global scenarios developed by the UN). In participatory activities, these types of scenarios should be communicated to participants in a clear way. Useful tools in this regard are visualization techniques such as realistic photographs or maps (Shaw et al. 2009; Sheppard et al. 2011) of the neighbourhood, city or farmland under alternative plausible futures (see Tool 3.1). Meteorological officers can support the facilitation of the session by explaining the data and joining group discussions (Wardekker et al. 2020b). Quantitative scenarios are particularly useful for experts, for instance, spatial planners who are to develop a sewer system in a climate-proof way.

- Option 2: Exploratory scenarios based on qualitative data

Futures can also be explored using more qualitative data. In an incremental scenario case study in the Golf of Morbihan, France (Wardekker et al. 2020b), organizers used poker design cards (see Tool 3.2) for participants themselves to actively and creatively explore various possible futures.

In this toolbox, the general objective of involving citizens in participatory foresight activities is to stimulate thinking about climate-resilient, hence more desirable futures. We therefore recommend not to use exploratory scenarios as a stand-alone method, but to add a normative element with other foresight methods, such as visioning, backcasting or seeds-based pathways.

SUMMARY **EXPLORATORY SCENARIOS** TOOLS

Tool 3.1 Visual maps (see Appendix A for more information)

> To visualize the impacts of global or national climate scenarios to the local context

Tool 3.2 Poker design cards (see Appendix A for more information)

> To let participants creatively develop possible local futures

3.2 Visioning

3.2.1 How to Use Visioning in a Participatory Activity?

- **Step 1: Imagine a Desired Future**

In visioning activities, participants are to envision their desired climate-resilient living environment (e.g. city, village, neighbourhood, or farmland) in the future. For visioning activities in the urban area, a list of climate adaptation measures can be consulted as an inspiration source (see Appendix B) (Runhaar et al. 2012).

Organizers can choose to let participants develop visions of their desired future freely, or with guidance.

- Option 1: Freely!

Visioning 'freely' means to develop visions of the future based on pure imagination. Creative tools such as Predict Future Headlines (Tool 3.4) and Creative collage (Tool 3.5) can help liberate the mind from present-day constraints by having participants use all their senses (see Fig. 3.2) (Carlsson-Kanyama et al. 2008; Raudsepp-Hearne et al. 2019; Pearson et al. 2018; Wardekker et al. 2020b).

Fig. 3.2 Creative material for Creative collage

- Option 2: A bit more guidance

– With questions about the past

Some participants may need some support to envision a radically different future neighbourhood, city or village. Since people tend to think about the future based on memories and experiences from the past, organizers could ask *'What are the things you love the most about [...] your community that you hope will still be there in the future?'* (Falardeau et al. 2019, p. 209). There may also be problematic elements in the present that citizens would like to see change in the future, such as experiences with extreme weather events (e.g. pluvial flooding, heat stress). Organizers can also make use of Photovoice (Tool 3.6); a visualisation tool that asks participants to make pictures of local places or issues that are personally relevant (Quinn and De Vrieze 2019). These photos can be further discussed and elaborated upon during the visioning activity.

– With existing good examples

Additional inspiration for a desired future can come from existing climate adaptation initiatives. Participants can then identify certain elements, skills or objectives of these *seeds* or 'pockets of the future' that they would like to see in their own context too (see Fig. 3.3) (Bennett et al. 2016; Falardeau et al. 2019; Hebinck et al. 2018). Participants can work with initiatives that they know of, but organizers can also prepare examples, for instance, from databases of local sustainability practices such as 'seeds of a good Anthropocene' (Bennett et al. 2016). The Future Wheels framework (Tool 3.7) helps participants explore how combinations of seeds can flourish in their neighbourhood, city or farmland (Falardeau et al. 2019).

Fig. 3.3 Seeds-based pathways

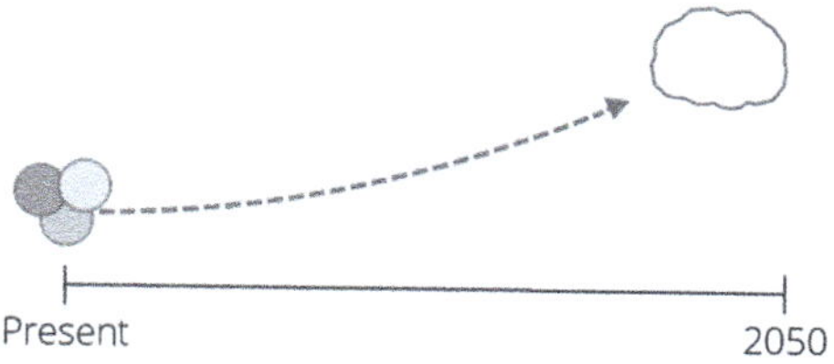

– With citizen narratives

In the preparation phase, we discussed how citizen narratives collected with interviews, surveys or focus groups can form the basis of visioning activities. They can, for instance, be turned into dimension cards (Tool 3.3) to serve as inspiration source for the visioning activity.

3.2.2 *Output Examples*

At the very least, a visioning activity gives people an experience of conscious engagement with the future. At the same time, visioning can generate various *products*, such as **collages** (see Fig. 3.4), **art designs** (see Fig. 3.5), or **poetry/spoken word**.

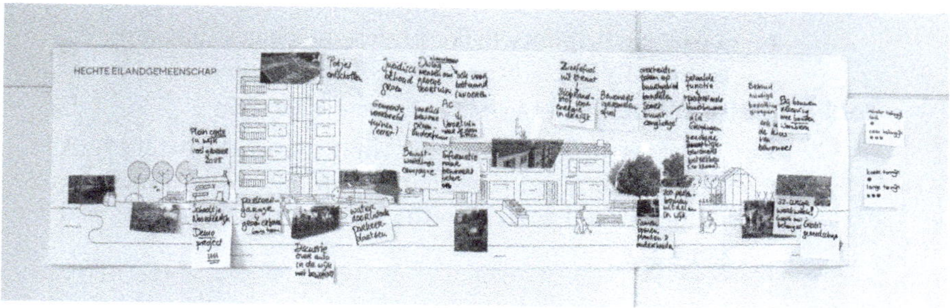

Fig. 3.4 Product of the visioning activity: collage. (From Wardekker et al. 2020b)

Fig. 3.5 Product of a visioning activity: art design. (From Pollastri et al. 2017)

SUMMARY **VISIONING** TOOLS

Tool 3.3 Dimension cards (see Appendix A for more information)

> To let participants 'flesh out' visions of the future with inspiration cards

Tool 3.4 Predict Future Headlines (see Appendix A for more information)

> To let participants creatively develop visions of the future

Tool 3.5 Creative collage (see Appendix A for more information)

> To let participants creatively develop visions of the future

> To facilitate interaction and collaboration between citizens and policymakers

Tool 3.6 Photovoice (see Appendix A for more information)

> To let participants make photos of places that are personally relevant as a basis for visioning

Tool 3.7 Future Wheels (see Appendix A for more information)

> To let participants imagine what the world would look like when a local sustainability initiatives becomes dominantand to explorethe wider impacts on the local context

3.3 Backcasting

3.3.1 *How to Use Backcasting in a Participatory Activity?*

- **Step 1: Start with the visioning output**

Backcasting activities start with one or multiple visions of a climate-resilient future (this can be the output of a visioning activity). In small groups, participants select one vision to work with. Especially when this vision is not developed by the participants, it is recommended to have a group discussion first about the legitimacy and relevance of the vision. The different groups can work with visions that seem mutually exclusive or unrealistic. In fact, when pathways are developed to completely different visions, it is interesting to explore whether they overlap, or how they might compete or constitute an obstacle (i.e. hinge-point) to each other (Wardekker et al. 2020b).

- **Step 2: Formulate actions**

The next step is to formulate (policy) actions, measures (see Appendix B for a list of examples), or interventions that participants deem essential to achieve their vision (Carlsson-Kanyama et al. 2008; Kok et al. 2011; Phdungsilp 2011). To ensure that participants formulate concrete actions, Kok et al. (2011, p.840) recommend using guiding questions of "*Why, How, When, How long, What, and Who?*". Some additional support of facilitators may be required here. If the goal of citizen participation in backcasting activities relates to policy development, it may be good to involve a select group of more experienced citizens.

Post-its and other creative material can be used to write down actions (see Figs. 3.6 and 3.7).

Fig. 3.6 Actions on post-its. (©Studio Lakmoes)

Fig. 3.7 Actions on post-its added to the vision. (©Studio Lakmoes)

- **Step 3: Label and order actions**

Group discussions about how to reach visions often result in a long list of suggested actions. The next step is to structure these ideas by categorizing them on their degree of importance; some actions are very crucial, whereas others are less necessary. An example of a labelling scheme used by Wardekker et al. (2020b), is: one green dot for less important actions, two dots for moderately important actions, and three dots for essential actions. Actions and their assigned label can be written on post-its (see Fig. 3.8).

Participants should also consider the time frame; some actions need implementation in the short-term (0–10 years), while others in the medium (10–20 years) or long-term (20–40 years). During this step, participants may realize that for some (important) actions to succeed, other actions need to precede. Such moments of reiteration are important for learning processes of both citizens and policymakers.

Fig. 3.8 Labelled actions. (©Studio Lakmoes)

- **Step 4: Discover storylines**

When all important actions are placed in the timeline, the final step is to identify combinations of actions around a common theme. For example, there may be actions related to social cohesion, health, or green innovation. By categorizing actions along themes, participants can formulate thematic storylines (i.e. pathways) to achieve their vision of the future. Wardekker et al. (2020b) formulated a 'social' pathway, a 'green community' pathway and a 'decentralized' pathway to a climate-resilient neighbourhood with strong social cohesion (see Fig. 3.9).

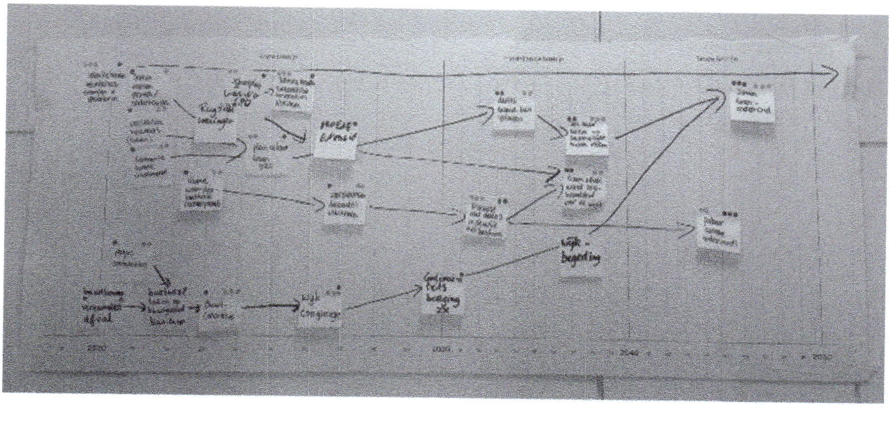

Fig. 3.9 Pathways. (From Wardekker et al. 2020a)

3.3.2 Output Examples

The output of backcasting activities can be a **timeline** with a sequence of actions written on post-its. The output can also be visualized by a professional artist. For instance, in a workshop organized by Hebinck et al. (2013), the thematic pathway of 'food provisioning and health' was visualized in a **cartoon**, with actions such as "Housing corporations have gardening 'coaches' that advise people on how to 'green' and maintain their neighbourhood", and "People and environment will be further stimulated to green their surrounding (e.g. through a subsidy)" (p. 17) (see Fig. 3.10).

Fig. 3.10 Art design of a backcasting pathway. (From Hebinck et al. 2013)

3.4 Incremental Backcasting

3.4.1 How to Use Incremental Backcasting in a Participatory Activity?

- **Step 1: Start with backcasting pathways on a timeline**

The first step is to take backasting pathways in the form of a timeline as the basis for incremental backcasting pathways (Wardekker et al. 2020b).

- **Step 2: Imagine what (unexpected) may happen on the way**

Next, participants brainstorm about examples of positive turns or negative constraints (i.e. hinge-points) that could influence their backcasting pathway. Hinge-points can be internal and controllable (e.g. the construction of a new sewage system), or external and uncontrollable (e.g. an economic crisis or a flood) (Wardekker et al. 2020a). Participants can place these events on the backcasting timeline. Since hinge-points can disrupt the pathway of actions, participants accordingly think about necessary steps or actions to address these disruptions and to still achieve the envisioned future. Participants can develop these alternative pathways of actions on the same backcasting timeline (Wardekker et al. 2020b).

3.4.2 Output Examples

The output of an incremental backcasting activity is the **timeline** that resulted from the backcasting activity complemented with hinge-points and alternative pathways (see Fig. 3.11) (Wardekker et al. 2020b).

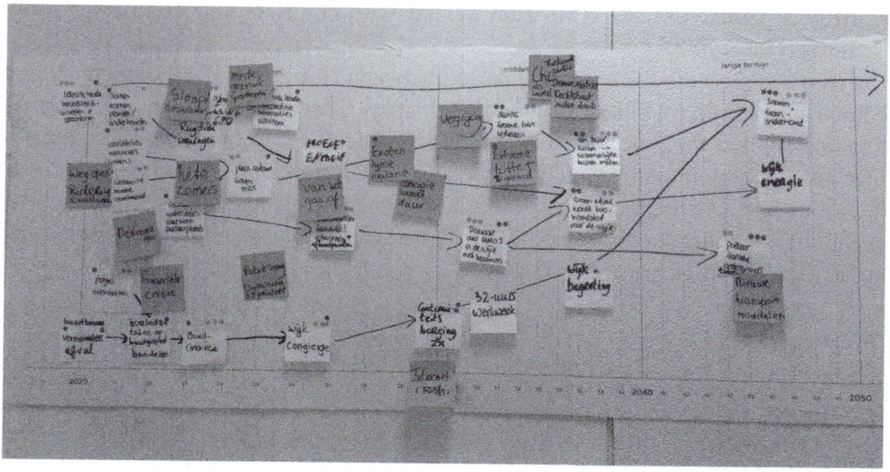

Fig. 3.11 The pink post-its indicate hinge-points and alternative pathways. (From Wardekker et al. 2020b)

3.5 Seeds-Based Pathways

3.5.1 How to Use Seeds-Based Pathways in a Participatory Activity?

Seeds-based pathways can be developed using the Three Horizons framework (Tool 3.8).

- **Step 1: Describe the present with seeds and maladaptation (Horizon 1)**

The first horizon refers to the present. Here, participants describe elements of the present that need to grow (seeds) or decline (maladaptation) to reach a climate-resilient future. The seeds related to climate adaptation should represent what participants would like to see happening in their climate-resilient neighbourhood, city or farmland. Participants can work with initiatives that they know of, or organizers can prepare examples. Participants can also be inspired by databases of local sustainability practices, such as the 'seeds of a good Anthropocene' database (Bennett et al. 2016).

Examples of maladaptation are a conservative culture in local spatial planning, common preference for paved gardens, or a lack of subsidy.

In the first horizon, the maladaptive practices are still dominant, while the seeds operate in the margins with minor impact on the neighbourhood or city as a whole (Raudsepp-Hearne et al. 2019).

- **Step 2: Envision a better future with mature seeds (Horizon 3)**

Horizon 3 represents a future situation in which seeds have grown into their mature form, while maladaptive practices, habits or policies have declined. Participants can be asked the question of '*What would the future look like if the combination of climate-adaptive practices becomes dominant?*' (Raudsepp-Hearne et al. 2019).

- **Step 3: Construct the transition period with conflicts and enabling conditions (Horizon 2)**

Horizon 2 represents the transition period between Horizon 1 and 3, which always comes with conflicting interests and disturbances. There may also be enabling factors that support the process of change (e.g. banning of cars from the urban centre; investments in community initiatives; rise of green political parties). For Horizon 2, participants can explore what is needed for positive change to happen. They can even look for connections with other seeds-based pathways to explore possible synergies (Raudsepp-Hearne et al. 2019).

3.5.2 Output Examples

Elements of the Three Horizons can be listed in a simple **table** (see Fig. 3.12) or visualized on a **timeline** (see Fig. 3.13). If more resources are available, a graphic designer can be invited to visualize the pathways in an **art form** (see Fig. 3.14).

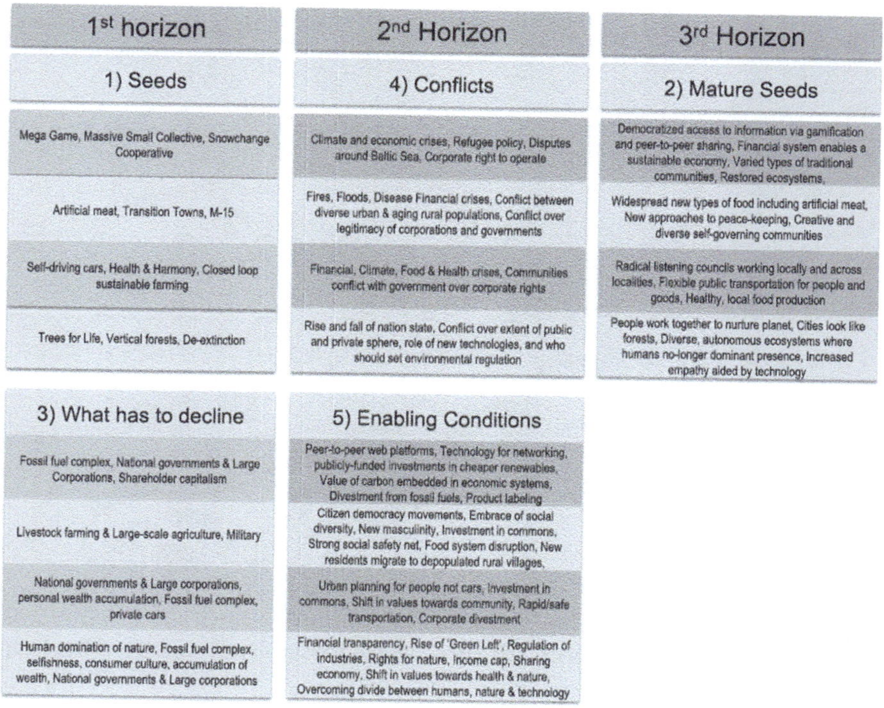

Fig. 3.12 Example elements of the Three Horizons framework. (From Raudsepp-Hearne et al. 2019)

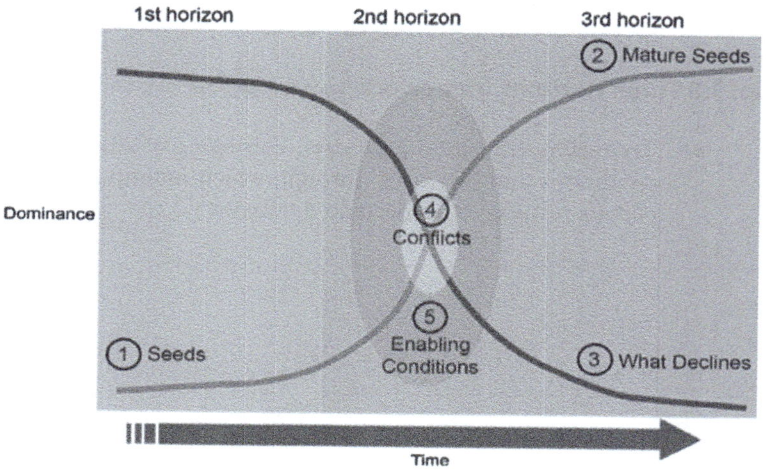

Fig. 3.13 The Three Horizons in the form of a timeline. (From Raudsepp-Hearne et al. 2019)

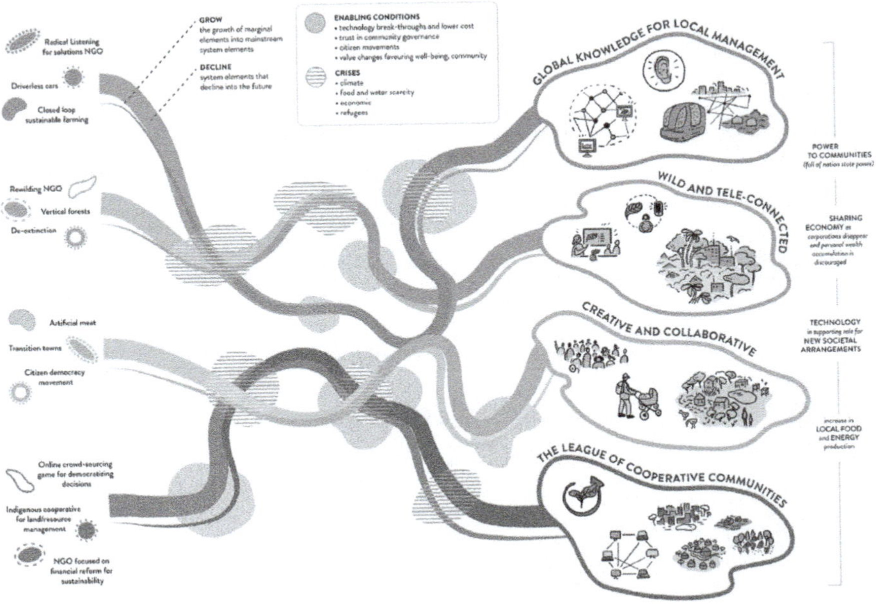

Fig. 3.14 Artistic visualization of the Three Horizons (i.e. transition) pathways. (From Raudsepp-Hearne et al. 2019)

SUMMARY **SEEDS-BASED PATHWAYS** TOOLS

Tool 3.8 Three Horizons framework (see Appendix A for more information)

> To explore transition pathways, consisting of enabling conditions and obstacles, through which initiatives related to climate-resilience can become dominant

Appendix A

Additional Information about Tools for Performing Participatory Foresight Activities

Tool 3.1: Visual Maps

- LEVEL ORGANIZER: ■■■
- LEVEL PARTICIPANT: ■□□
- DURATION: 30–60 mins

What? Visual GIS maps (e.g. 2D, 3D, fly-over) of the local living environment under different (climate) scenarios can let people imagine possible future situations. These maps could contain any information about climate change and how it affects the local area, from flooding to extreme heat. Visual maps make use of people's emotional connection to the place and as such bring climate change to life (Sheppard et al. 2011).

When? In participatory activities with other foresight methods such as visioning, backcasting, or seeds-based pathways.

How? Maps of the local area under different scenarios can be shortly presented by the organizer or a (meteorological) expert at the beginning of the activity. Experience shows that the involvement of an expert in group discussions enhances understanding among participants as well as improves the credibility of scenario trends (Wardekker et al. 2020b). Furthermore, experts often enhance trust building by potential end-users of climate services such as citizens and policymakers. The chance that they will use such climate information in the future thereby increases.

A visual map showing a neighbourhood in Dordrecht under the impact of extreme precipitation events. From Wardekker et al. (2020b)

→See Wardekker et al. (2020b), Sheppard et al. (2011), and Shaw et al. (2009) for more information about visual maps

Tool 3.2: Poker Design Cards

- LEVEL ORGANIZER: ■■□
- LEVEL PARTICIPANT: ■□□
- DURATION: 30–60 mins

What? Poker design cards help participants think out of the box by exploring different futures based on randomly selected inspiration cards.

When? In participatory activities with other foresight methods such as visioning, backcasting, or seeds-based pathways.

How? The cards used by Wardekker et al. (2020b) in the workshop in the Gulf of Morbihan, France, contained elements that according to citizens could potentially have local impact. The narratives were divided into three categories: 'Climate change and hazards'; 'Infrastructure and territory'; and 'Resources and actors'. Participants were asked to randomly pick one card of each category and use these three to describe a possible future situation.

Examples of Poker design cards. From Wardekker et al. (2020b)

Poker design categories	Categories of narratives			
	Geo-social	Historical	Seasonal	Climatic effects
Climate change and hazards	Submersion	Drying soils	Warmer summer and spring periods	Storms
	Flooding	Sea level rise		Heat waves
	Erosion	Ocean acidification	Colder winters	Droughts
Infrastructure and territory	First nautical mile	Oyster farms	Second homes	Historical sites
	Subsidence	Coastal pathway	Ports	Urban areas
	Beaches	Salt mines	Water treatment systems	Routes
Resources and actors	Island owner	Oyster farmers	Office of Tourism	Measuring instruments
	Intra-gulf nautical transport network	Direct selling	Retired population	Scientific community
		Tourists	Seasonal workers	

A participant using poker design cards. From Wardekker et al. (2020b)

Tool 3.3: Dimension Cards

- LEVEL ORGANIZER: ■■□
- LEVEL PARTICIPANT: ■□□
- DURATION: 30–60 mins

What? Dimension cards (e.g. based on citizen narratives) guide and inspire participants when envisioning a desired future.

When? In participatory visioning activities.

How? Wardekker et al. (2020b) focused on the goal of a climate-resilient Bergen (Norway) in 2050. The visioning activity started with randomly allocating participants to one of three broad visions of Bergen in 2050: 'Control the climate' (*a 1.5 degree city*), 'Live with the climate' (*let it rain*), or 'Make the most of the climate' (*high-tech haven*). These visions were prepared beforehand based on interviews with citizens. During the visioning activity, participants added more detail using the dimension cards with elements that, according to the interviewees, give Bergen a sense of place. Participants voted for five cards to use as the basis for their vision.

1 A compact city	**5** A climate science city	**9** Freeing the waterways	**13** A city linked to nature
2 climate-proof buildings	**6** resilient Bergensers	**10** safe from climate impacts	**14** diverse and international
3 A port city	**7** A historical city	**11** rain-friendly spaces in the city	**15** green spaces in the city
4 walkways and cycle-ways	**8** A local democracy	**12** busses, boats	**16** blank card

Example of dimension cards. From Wardekker et al. (2020b)

Tool 3.4: Predict Future Headlines

- LEVEL ORGANIZER: ■■□
- LEVEL PARTICIPANT: ■□□
- DURATION: 30–60 mins

What? Predict Future Headlines is an accessible visioning tool that invites participants to time-travel to a celebrative future moment in which the local living environment has turned into a climate-resilient place and to visualize this desired vision with a creative newspaper headline (Pearson et al. 2018).

When? In participatory visioning activities.

How? Participants develop an imaginary newspaper headline as if it was that moment in time. They can use creative material from journals or magazines.

From Pearson et al. (2018)

Tool 3.5: Creative Collage

- LEVEL ORGANIZER: ■■□
- LEVEL PARTICIPANT: ■■□
- DURATION: 60 mins

What? Creative collage is a visioning tool to let participants develop a desired future with creative materials. The tool works well in group activities as it is easy to understand and allows less outspoken participants to share their ideas too (Pearson et al. 2018).

When? In participatory visioning activities.

How? Before the activity, Wardekker et al. (2020b) identified three thematic visions of a future neighbourhood based on citizen narratives collected earlier: 'Water safe', 'Community-oriented' and 'Innovation-oriented'. Graphic designers visualized these themes in templates. During the workshop, participants elaborated on these broad visions using creative material on the templates. Some out-of-the-box images were provided to invite participants to go beyond their customary reasoning and to stimulate innovative ideas.

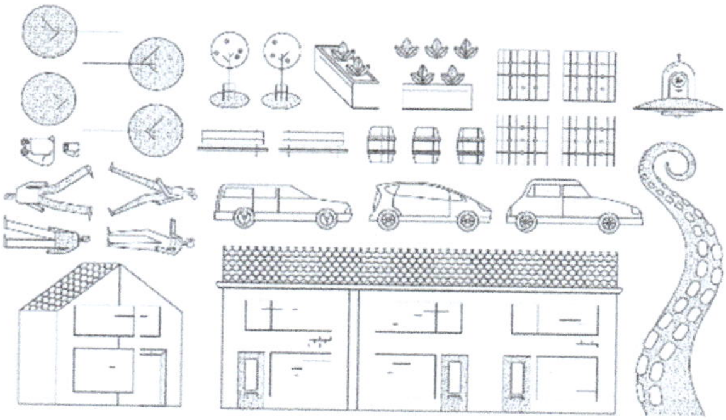

Creative material for collage. ©Studio Lakmoes

Tool 3.6: Photovoice

- LEVEL ORGANIZER: ■■□
- LEVEL PARTICIPANT: ■□□
- DURATION: 30–60 mins

What? Photovoice is a tool to let participants capture places or issues in their living environment that are meaningful to them or that they would like to change in the future (Quinn and De Vrieze 2019).

When? In participatory visioning activities.

How? Participants take the pictures prior to the visioning activity. During the activity, they discuss and share their photos with fellow citizens and local policymakers. Photovoice activities range from short discussions (e.g. as an introduction to the foresight activity) to longer sessions (e.g. as the main activity).

Pictures of the Vogelbuurt neighbourhood. From Wardekker et al. (2020b)

Tool 3.7: Future Wheels

- LEVEL ORGANIZER: ■■■
- LEVEL PARTICIPANT: ■■■
- DURATION: 2+ h

What? With the Future Wheels tool participants can envision the consequences of a climate-resilient future on their local living environment (Falardeau et al. 2019; Pereira et al. 2018; Raudsepp-Hearne et al. 2019).

When? In participatory visioning activities.

How? First, the participants choose one or more local initiatives related to climate adaptation (e.g. from the 'seeds of a good Anthropocene database, see Bennett et al. (2016)) and place them on a blank sheet. Each seed forms the center of one future wheel. Participants then discuss possible (direct and indirect) effects of the mature/dominant version of the seed on the local context. The further away in the future, the further away effects are placed from the center. The STEEP approach (Social, Technological, Economic, Environmental and Political impacts) can be used as a checklist to ensure that all types of effects are covered (Falardeau et al. 2019; Pereira et al. 2018; Raudsepp-Hearne et al. 2019).

Future Wheels. From Raudsepp-Hearne et al. (2019)

Tool 3.8: Three Horizons Framework

- LEVEL ORGANIZER: ■■■
- LEVEL PARTICIPANT: ■■■
- DURATION: 2+ h

What? With the Three Horizons Framework, participants can explore transition pathways for local climate adaptation initiatives. The term 'Three Horizons' refers to the three phases that these initiatives need to go through to scale up and become mainstream and the new status quo (1 = business-as-usual phase, 2 = transition phase, 3 = sustainable state). The tool helps identify the system characteristics that need to change for a climate-resilient future (Raudsepp-Hearne et al. 2019).

When? In participatory activities with seeds-based pathways.

How? First, each group describes the present by identifying three existing local initiatives (i.e. 'seeds') in relation to climate adaptation, as well as some maladaptive practices (Horizon 1). Examples of seeds are 'vertical forests' (i.e. vertical densification of nature in the city that supports naturalization of large urban and metropolitan borders) and 'transition towns' (i.e. grassroots community projects in response to climate change and economic instability). Second, participants describe the seeds in their mature form (Horizon 3). Finally, they identify enabling conditions for the seeds to grow from the margins to a dominant state, as well as possible obstacles that can be encountered (Raudsepp-Hearne et al. 2019).

Appendix B: Example List of Climate Adaptation Measures in Urban Areas

From Runhaar et al. (2012).

Climate impact to adapt to	Timing of measures	Measures
Heat stress	Proactive	Open water, fountains
		Vegetation (cooling due to evaporation)
		High albedo pavement instead of asphalt
		Creating optimal shading in building orientation, compact building and (big leaf) trees
		Orientation and profile of streets regarding wind direction (affecting wind speed and urban ventilation)
		Replacement of vulnerable groups
		Monitoring and inspection
		Warning systems and disaster contingency plans
	Reactive	Wetting streets and roofs

(continued)

Climate impact to adapt to	Timing of measures	Measures
Flooding	Proactive	Seeping water 'screens'
		Water permeable pavement instead of asphalt and other measures for better infiltration and water outlet
		Lower water tables
		Separation of rainwater and sewage water plumbing
		Enhancing capacity of sluices and weirs
		Elevate urban areas
		Additional flood defences (dykes or buildings) or reinforcing existing ones
		Replacement of vulnerable buildings and infrastructure
		Disaster contingency plans (e.g. temporary dykes)
		Monitoring and inspection
		Warning systems
		Evacuation plans
		Extra green space
		Water storage facilities (open water such as pools)
		Increase sewer capacity or enhanced maintenance
		Drainage systems
		Dry pumps and other provisions for water discharge and clean-up
	Reactive	Warnings and information
		Clean-up and damage remedy
		Recovery plans

References

Bennett EM, Solan M, Biggs R, McPhearson T, Norström AV, Olsson P, Pereira L, Peterson GD, Raudsepp-Hearne C, Biermann F, Carpenter SR, Ellis E, Hichert T, Galaz V, Lahsen M, Milkoreit M, Martin López B, Nicholas KA, Preiser R et al (2016) Bright spots: seeds of a good Anthropocene. Front Ecol Environ 14(8):441–448. https://doi.org/10.1002/fee.1309

CARE International (2018) Practical guide to participatory scenario planning: seasonal climate information for resilient decision-making. CARE International, Geneva

Carlsson-Kanyama A, Dreborg KH, Moll HC, Padovan D (2008) Participative backcasting: a tool for involving stakeholders in local sustainability planning. Futures 40:34–46. https://doi.org/10.1016/j.futures.2007.06.001

Crawford MM (2019) A comprehensive scenario intervention typology. Technol Forecast Social Change 149:119748

Dammers E, Van 't Klooster S, De Wit B, Hilderink H, Petersen A, Tuinstra W (2013) Scenario's maken voor milieu, natuur en ruimte: een handreiking. PBL, Den Haag

Falardeau M, Raudsepp-Hearne C, Bennett EM (2019) A novel approach for co-producing positive scenarios that explore agency: case study from the Canadian Arctic. Sustain Sci 14:205–220. https://doi.org/10.1007/s11625-018-0620-z

Hebinck A, Villarreal G, Oostindie H, Hebinck P, Zwart TA, Vervoort J, Rutting L, & De Vrieze A (2013) Urban agriculture policy-making: Proeftuin040 – TRANSMANGO scenario workshop report, the Netherlands

Hebinck A, Vervoort JM, Hebinck P, Rutting L, Galli F (2018) Imagining transformative futures: participatory foresight for food systems. Ecol Soc 23(2):16

IPCC (2018) Summary for policymakers. In: Global warming of 1.5°C. An IPCC special report on the impacts of global warming of 1.5°C above pre-industrial levels and related global greenhouse gas emission pathways, in the context of strengthening the global response to. https://www.ipcc.ch/

Kok K, Van Vliet M, Bärlund I, Dubel A, Sendzimir J (2011) Combining participative backcasting and exploratory scenario development: experiences from the SCENES project. Technol Forecast Social Change 78:835–851. https://doi.org/10.1016/j.techfore.2011.01.004

Lorenz S, Dessai S, Forster PM, Paavola J (2017) Adaptation planning and the use of climate change projections in local government in England and Germany. Reg Environ Chang 17:425–435. https://doi.org/10.1007/s10113-016-1030-3

Pearson KR, Bäckman M, Grenni S, Moriggi A, Pisters S, De Vrieze A (2018) Arts-based methods for transformative engagement: A toolkit. https://doi.org/10.18174/441523

Pereira L, Hichert T, Hamann M, Preiser R, Biggs R (2018) Using futures methods to create transformative spaces: visions of a good Anthropocene in southern Africa. Ecol Soc 23(1). https://doi.org/10.5751/ES-09907-230119

Phdungsilp A (2011) Futures studies' backcasting method used for strategic sustainable city planning. Futures 43:707–714. https://doi.org/10.1016/j.futures.2011.05.012

Pollastri S, Boyko C, Cooper R, Dunn N, Clune S, Coulton C (2017) Envisioning urban futures: from narratives to composites. Des J 20:S4365–S4377. https://doi.org/10.1080/14606925.2017.1352933

Quinn M, De Vrieze A (2019) Creating sustainable places together. A quick start guide for policymakers and practitioners to place-based working and co-production

Raudsepp-Hearne C, Peterson GD, Bennett EM, Biggs R, Norström AV, Pereira L, Vervoort J, Iwaniec DM, McPhearson T, Olsson P, Hichert T, Falardeau M, Jiménez Aceituno A (2019) Seeds of good anthropocenes: developing sustainability scenarios for northern Europe. Sustain Sci. https://doi.org/10.1007/s11625-019-00714-8

Robinson J (2003) Future subjunctive: backcasting as social learning. Futures 35:839–856. https://doi.org/10.1016/S0016-3287(03)00039-9

Runhaar H, Mees H, Wardekker A, Van der Sluijs J, Driessen P (2012) Adaptation to climate change-related risks in Dutch urban areas: stimuli and barriers. Reg Environ Chang 12:777–790. https://doi.org/10.1007/s10113-012-0292-7

Shaw A, Sheppard S, Burch S, Flanders D, Wiek A, Carmichael J, Robinson J, Cohen S (2009) Making local futures tangible—synthesizing, downscaling, and visualizing climate change scenarios for participatory capacity building. Glob Environ Chang 19:447–463. https://doi.org/10.1016/j.gloenvcha.2009.04.002

Sheppard SRJ, Shaw A, Flanders D, Burch S, Wiek A, Carmichael J, Robinson J, Cohen S (2011) Future visioning of local climate change: A framework for community engagement and planning with scenarios and visualisation. Futures 43:400–412. https://doi.org/10.1016/j.futures.2011.01.009

UN Environment (2019) Global environment outlook – GEO-6: healthy planet. Healthy People. https://doi.org/10.1017/9781108627146

Van Notten PWF, Rotmans J, Van Asselt MBA, Rothman DS (2003) An updated scenario typology. Futures 35:423–443. https://doi.org/10.1016/S0016-3287(02)00090-3

Wardekker A, Bremer S, Krauß W, Da Cunha C, Paula Farias Rocha A, Baztan J, Jaffrès L, Breton F, Runhaar H, Vervoort J, Van der Sluijs J, Wildschut D, Vanderlinden J-P (2020a) Protocol for designing incremental scenarios. CoCliServ, Guyancourt, France.

Wardekker A, Van den Ende M, Marschutz B, Pijnappels M, Hofland S, Bremer S, Blanchard A, Iversen L, Van der Sluis J, Krauß W, Rocha A, Da Cunha C, Baztan J, Jaffrès L (2020b) Incremental scenario case studies. CoCliServ, Guyancourt, France.

Chapter 4
Reflection: Applying Participatory Foresight Methods in Practice

Abstract In this final chapter of the toolbox, we present two illustrative cases of participatory activities with foresight methods: one organized in an urban context in the Netherlands, and one in a rural context in Honduras. Both cases are discussed in detail following the structure of this toolbox; we start with exploring the questions of *why citizen participation* was chosen, *who were involved* in the activity, *which foresight methods* were used (from Chap. 2) and *how these methods were used* (from Chap. 3).

Keywords Toolbox · Citizen participation · Futures · Foresight methods · Scenarios · Visioning · Pathways · Resilience · Climate adaptation · Urban · Rural · Sustainability

4.1 A Participatory Foresight Workshop in Dordrecht, the Netherlands (Urban Context)

The city of Dordrecht, the Netherlands, is highly vulnerable to weather and climate change due to its geographical location: it lies in an urban delta under sea-level, close to the sea, and surrounded by rivers. Furthermore, there are several deteriorated neighbourhoods, such as 'de Vogelbuurt' that require spatial redevelopment (see Fig. 4.1). In this light, the municipality formulated an ambition of a climate-resilient city in 2040 and thereby recognized an important role for citizens. The question remained: what would a climate-resilient Dordrecht look like according to local stakeholders and what are ways to reach it? In this context, Wardekker et al. (2020) organized a foresight workshop for citizens, policymakers and scientists.

M. A. van den Ende et al., *Towards a Climate-Resilient Future Together*, SpringerBriefs in Environmental Science,
https://doi.org/10.1007/978-3-031-07682-4_4

Fig. 4.1 Neighbourhood 'de Vogelbuurt', Dordrecht. (From Wardekker et al. 2020)

4.1.1 Why Citizen Participation?

The goal of the foresight workshop was to identify the desired climate-resilient future of people living in neighbourhood 'de Vogelbuurt' and how they could reach it. In particular, organizers were keen to identify citizens' need for new climate services (i.e. the knowledge and capacity building goal) (Wardekker et al. 2020). Citizens also had to be involved in order to include their desires and fears in the neighbourhood adaptation plan (i.e. the policy development goal) as well as to simply provide a space for the local municipality to engage on the topic of climate adaptation (i.e. the community building goal).

4.1.2 Who to Involve?

Relevant stakeholders in this case study were residents of neighbourhood 'de Vogelbuurt' in Dordrecht, local policymakers and scientists. Prior to the workshop, scientists went into the neighbourhood to collect citizens' stories through informal talks, interviews and focus groups (Marschütz et al. 2020). As such, the scientists collected some first ideas of the local community on how to make the neighbourhood climate proof. The scientists' regular appearance in the neighbourhood also enhanced trust-building, which was expected to increase the willingness of citizens to attend the participatory foresight workshop.

The local community centre promoted the workshop and citizens were also contacted by the researchers. Nevertheless, the participation rate was rather low. Two community workers who were present at the workshop raised that the duration of the workshop (a full day) as well as the location (in another neighbourhood) discouraged citizens to attend. Citizens also indicated their distrust in the municipality.

Post-workshop Reflection on the Goal of Participation and the Low Participation Rate

Some of the workshops discussed in Wardekker et al. (2020), including the one in Dordrecht, had a relatively low participation rate. Especially for participatory activities with a policy development goal, a larger group of citizens is crucial to make sure that a diversity of perspectives is represented in policymaking. Therefore, they advise to supplement longer workshops with shorter evening sessions in the neighbourhood to involve more citizens and other non-professional actors in developing practical follow-up measures. In addition, to reach the knowledge and capacity building goal, follow-up workshops are planned to be organized with the producers and users of climate services: climate experts, local governments and community representatives.

4.1.3 Which Foresight Methods to Use? And How to Use Foresight Methods?

4.1.3.1 Exploratory Scenarios

Wardekker et al. (2020) used exploratory climate scenarios in several ways throughout the workshop. First, a national meteorological expert presented a publicly available database[1] with historic, present and future climate data that was downscaled to the local level. After the presentation, the participants were asked: '*What could changing weather/climate mean for the neighbourhood in the future?*' A laptop was available with a spreadsheet of climate data and a page with 3D fly-over maps of the neighbourhood showing information about flood risks and heat stress under two different scenarios (see Fig. 4.2). Participants used both the climate data and the visual maps during the visioning activity. They also used it to identify possible hinge-points.

[1] See http://www.klimaatscenarios.nl/images/Brochure_KNMI14_NL.pdf and https://data.knmi.nl/datasets

Fig. 4.2 Visual map showing the impacts of heat stress. (From Wardekker et al. 2020)

4.1.3.2 Visioning

Before the workshop, Marschütz et al. (2020) had already collected narratives of citizens about how they would like to see their neighbourhood in the future under climate change. The result, a broad vision of a resilient island of Dordrecht, was used by Wardekker et al. (2020) to develop three sub-visions: (1) 'Strong island community', (2) 'Innovative connections', and (3) 'Water safe and water wise'.

During the workshop, each group of participants received an illustrated XL-card of a street created by graphic designers Studio Lakmoes (see Fig. 4.3) to visualize a climate-resilient neighbourhood with creative material. Post-its were used to write down actions and measures needed to achieve this vision. The visioning activity resulted in two collages of a climate-resilient neighbourhood with measures such as more green and trees, not only to reduce heat stress but also to strengthen community cohesion (see Fig. 4.4). At the same time, they wanted enough parking space. Participants also envisioned community gardens to take care of collectively. These ideas were used as a basis for backcasting pathways.

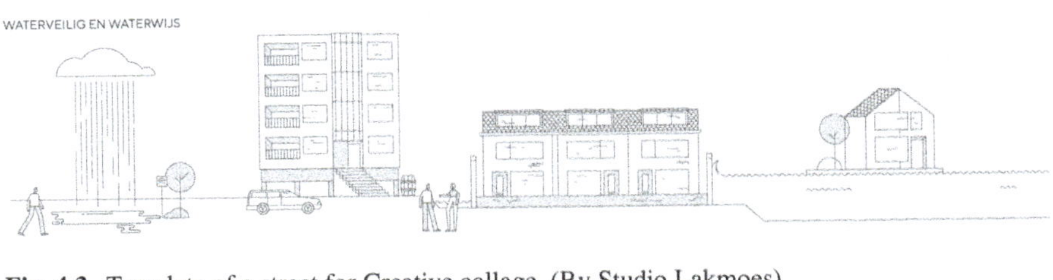

Fig. 4.3 Template of a street for Creative collage. (By Studio Lakmoes)

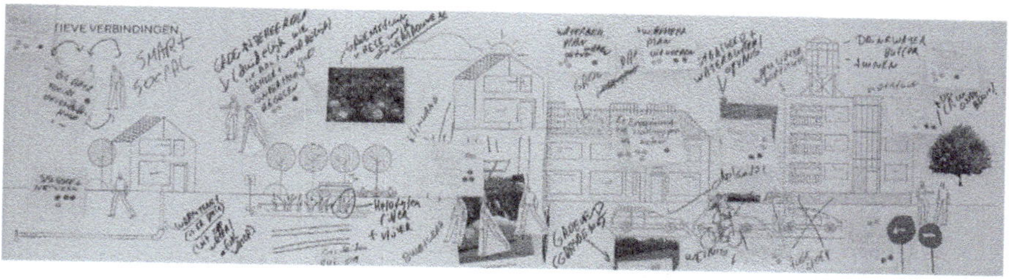

Fig. 4.4 Result of the visioning activity using creative collage. (From Wardekker et al. 2020)

4.1.3.3 Backcasting

During the visioning activity, participants identified several measures to achieve a climate-resilient neighbourhood. In the backcasting activity, these measures were written down on a second set of post-its. Accordingly, participants were asked to critically reflect on how essential these measures are for achieving their future neighbourhood, as well as think about when they should be implemented. Participants answered these questions for each measure and labelled them accordingly.

The next step was to place the measures on a timeline and identify routes or pathways of measures (see Fig. 4.5). The participants formulated a pathway of social measures and green measures.

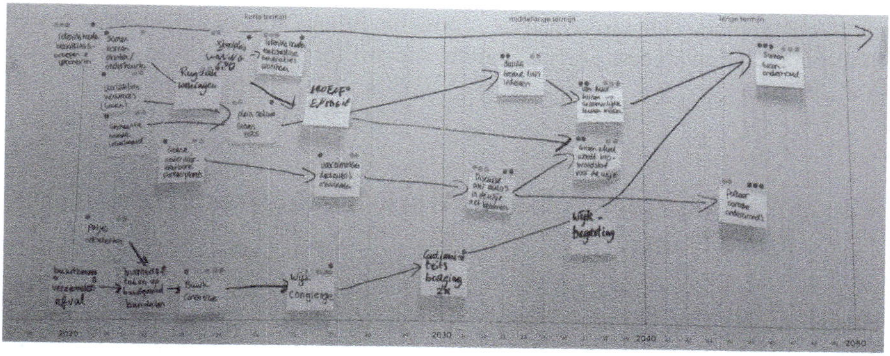

Fig. 4.5 Pathways of measures. (From Wardekker et al. 2020)

4.1.3.4 Incremental Backcasting

The group was then asked to formulate critical moments that could either support the implementation or reduce the effectiveness of measures (i.e. hinge-points). Participants identified the following hinge-points (some accelerating, others

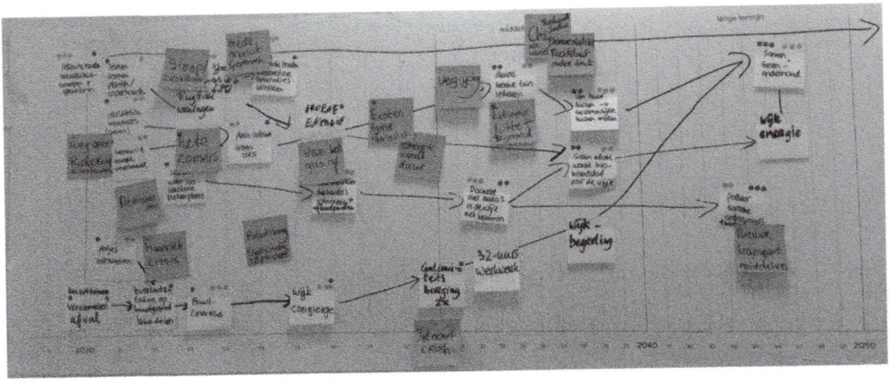

Fig. 4.6 Pathways and hinge points. (From Wardekker et al. 2020)

constraining the backcasting pathways): a financial crisis, the reconstruction of a street, expensive energy, and extreme weather events such as hot summers and extreme rainfall. Also a lack of community support and a lack of awareness about climate change impacts were identified as potential obstacles for the neighbourhood to become climate-resilient in 2050 (see Fig. 4.6).

In the final step, the scientists asked the participants which information they would need to anticipate or to overcome the obstacles, hence to succeed with the implementation of measures. One citizen proposed posters of the neighbourhood under different scenarios (e.g. one with trees and one without trees, or under different temperatures). This visualization of possible futures was a climate service need that could raise awareness and gain public support for adaptation plans (Wardekker et al. 2020).

4.2 A Participatory Foresight Workshop in Honduras (Rural Context)

Honduras is one of the most climate-affected countries in the world. Honduras' agricultural sector, the backbone of the economy, is extremely vulnerable to climate variability and change. Bad weather conditions and fungus from increasing temperatures are already tormenting farmers by crippling yields and profits from the previously successful coffee and banana plantations. To face these challenges, the government felt the need to work together with a wide range of different people and organizations to come up with transformative new strategies that lead to action.

In line with this thinking, a national workshop was held in one of the two most vulnerable regions of Honduras, Choluteca, to test and thoroughly revise a draft climate strategy by the Secretariat of Agriculture and Livestock using future scenarios. The "Strategy for risk management and climate change adaptation for the sector of agriculture and livestock" was tested against country-based socioeconomic and environmental scenarios combined with climate impacts on the agricultural sector (Schubert 2014). In her role as Scenarios Coordinator for Latin America for CCAFS, Marieke Veeger (scientist at the University for International Cooperation in Costa Rica) led the process. The University of Oxford's global scenario team led by Joost Vervoort offered methodological support.

4.2.1 *Why Citizen Participation?*

In this example, the need for wider participation was taken very seriously by the national government, who were interested in developing a national strategy that reflected the needs of all who might be hit by climate change (i.e. policy development goal). Because of this need for an integrated strategy for the rural sector, the scenario-guided policy formulation offered an opportunity for developing new links between people across different parts of society, and across different scales (i.e. community building goal); and for developing new capacities around scenario-guided planning in an integrated manner among all these societal groups (i.e. knowledge and capacity goal).

What was really unique about this process is that the government was extremely open to having their ideas about the national strategy challenged by diverse societal perspectives. Resultantly, many parts of the national strategy were deeply revised, and an entire section on longer-term adaptation was added before the strategy was finalized and accepted by the government. The Honduran President later commented that he 'could hear the voices of the farmers' in the final strategy. The following quote by Marieke Veeger illustrates this:

> 'I was really happy to see that the participants, some of whom had led the strategy work themselves, were so open to the suggested changes. That is not always a given in this work.'

4.2.2 *Who to Involve?*

Relevant stakeholders in this case were very diverse, including many different types of farmers (large-scale and small-scale farmers), people from various branches of the Honduran government, other private sector actors, civil society organizations, and academics. Despite initial disturbances in the organization because of personnel changes in the government, the process was able to bring all these people together. Key in doing this was the very strong relationship between CCAFS regional team (Ana Maria Loboguerrero as CCAFS Regional Program Leader and Deissy Martinez as Science Officer) and the Honduran government, since they had been working together on many projects. The Honduran government team, moreover, were uniquely open, flexible, and interested in truly inclusive participation and the use of new scenario methods. Marieke Veeger explained that:

> 'What's good about the Scenarios work is that it can really help policy-makers to strengthen a current plan or policy, without requiring too much time and effort from them. I believe that is what our partners find the most attractive.'

Because of this, the timing of the workshop was perfect – the new strategy had been developed into a basic draft, but everything was still up for change. Every participant knew they would get an opportunity to profoundly impact the strategy, and thereby, future government action. The workshop was organized in a rural centre, away from the capital, to put the participants in the heart of the action. Reflecting on the participation process, coordinator Marieke Veeger stated that (Schubert 2014):

> 'Success is many times dependent on good timing, something you might not be able to influence. But we have found that building the right relationships with key influential stakeholders and getting them to participate in the workshops is crucial in order for policies to change. Also, making sure that there is a plan or policy ready to be tested, which will be implemented regardless of the Scenario activities, is also key in order to achieve traction.'

4.2.3 *Which Foresight Methods to Use? And How to Use Foresight Methods?*

Contrary to the Dordrecht city planning example, in this case, the foresight activity was entirely developed around an on-going government strategy development activity. This means that there was already a draft vision on the table; it was the job of the participants to critique, expand and improve the vision and the steps to reaching it. Because of this, the workshop focused primarily on the use of exploratory scenarios as a testing tool for the strategy.

Before the scenario activity could start, on day 1 of this 2-days workshop, participants had to familiarize themselves with the ideas in the strategy. The group of around 40 participants was divided into thematic groups, each of which selected a part of the strategy to closely investigate. The thematic groups read the strategy and already provided a first round of comments and suggestions. This was helpful, because when the scenarios were used in the second round of reviews, it was very clear that they brought up very different recommendations depending on the scenario. Furthermore, this first assessment helped identify the key issues that the scenarios should address to make them relevant for the analysis of the strategy.

4.2.3.1 National Exploratory Scenarios

Another element that made this process different from the Dordrecht case is that the scenario development was based on pre-existing scenarios for the Central American region. These scenarios were developed by the CCAFS program, in another participatory foresight process with experts and stakeholders from across Central America (including, importantly, some of the people in this workshop in Honduras!). The Central American scenarios were very important because they offered a regional context for the Honduran situation, while maintaining a link with global trends. Some modelling work was available that offered quantitative scenario information about the future availability of crops, land and other aspects.

On day 2, four new groups were created, with a different mix of people from the different thematic groups. Each of these groups focused on one of the Central American scenarios, and considered, how would this scenario play out in Honduras? They first created a description of the Honduran version of the scenario for 2050. Then they used back-casting, not to set a goal but to create a reverse storyline to connect the scenario to the present. Finally, they used a list of all the issues considered important for the national strategy from the first day to flesh out the details of the scenario.

4.2.3.2 Policy Analysis Based on Exploratory Scenarios

By now, each of the four groups had deep familiarity with the down-scaled, Honduran scenario they had created. In a next step, each group received copies of the climate strategy, with additions made by the theme groups in the first day. Each of the scenario groups was asked to evaluate, in great detail, this entire new draft of the strategy from the perspective of their specific scenario. Would the different elements of the strategy work in this scenario? If not, why not? What aspects of the strategy were still too vague, not concrete enough, not thought out? What aspects of the strategy were most vulnerable? And importantly, how could these vulnerabilities and gaps be improved to make the plan more robust and actionable?

Because this analysis was happening from the perspectives of four different scenarios, each group came up with original, diverse insights to help improve the strategy. By discussing the different insights, common recommendations emerged, leading to fundamental expansions and improvements of the strategy. Marieke Veeger said that (Schubert 2014):

> 'It quickly became very clear that the strategy had to be diversified, and include other types of livelihoods, such as cattle and poultry businesses too. Participants also suggested to include territorial planning in its objectives to guarantee most fertile lands for agriculture, since several of the scenarios showed drastic urban expansion.'

In addition to the finalization of the national climate strategy, the successful collaboration and use of scenario planning between CCAFS, the Honduran government and a range of other organizations led to a number of new collaborations for other national strategies (Fig. 4.7).

Fig. 4.7 Art design of the scenario-based policy analysis in Honduras. (From Schubert 2014)

References

Marschütz B, Bremer S, Runhaar H, Hegger D, Mees H, Vervoort J, Wardekker A (2020) Local narratives of change as an entry point for building urban climate resilience. Clim Risk Manag 28:100223. https://doi.org/10.1016/j.crm.2020.100223

Schubert C (2014) Helping Honduras build a more robust climate adaptation strategy for the agriculture sector. https://ccafs.cgiar.org/news/helping-honduras-build-more-robust-climate-adaptation-strategy-agriculture-sector

Wardekker A, Van den Ende M, Marschutz B, Pijnappels M, Hofland S, Bremer S, Blanchard A, Iversen L, Van der Sluis J, Krauß W, Rocha A, Da Cunha C, Baztan J, Jaffrès L (2020) Incremental scenario case studies. CoCliServ, Guyancourt, France.